Uchuva (*Physalis peruviana* L.) Reproductive Biology

Fernando Ramírez • Thomas Lee Davenport

Uchuva (*Physalis peruviana* L.) Reproductive Biology

 Springer

Fernando Ramírez
Research
Independent Researcher
Bogotá, Colombia

Thomas Lee Davenport
Tropical Res. and Education Center
University of Florida
Homestead, FL, USA

ISBN 978-3-030-66554-8   ISBN 978-3-030-66552-4   (eBook)
https://doi.org/10.1007/978-3-030-66552-4

This Springer imprint is published by the registered company Springer Nature Switzerland AG
The registered company address is: Gewerbestrasse 11, 6330 Cham, Switzerland

The first author dedicates this book to his parents Natalia and Fernando Ramírez

Preface

Worldwide, uchuva (Cape gooseberry), *Physalis peruviana* L. has achieved great popularity over the years (Fig. 1). It is a delicious fruit to eat but also contains many health improving properties, such as cancer prevention, inhibition of tumor promotion, and prevention of age-related diseases. It is also antispasmodic, diuretic, and analgesic. Uchuva has gained popularity in numerous countries and is included as a garnish in many dishes. The plant is mainly cultivated in the tropics but has also gained entrance in the subtropics and under glass houses in temperate regions of the world.

The reproductive biology of the plant has been studied by different researchers worldwide, but a reference work is lacking. This book reviews the scientific literature describing the reproductive biology of uchuva. It includes recent publications by the authors about related Solanaceous species. The book brings together aspects that are essential for the understanding of this plant´s cultivation in tropical and subtropical conditions.

The book provides an in-depth view of uchuva reproduction. It includes information obtained from articles and texts describing uchuva horticulture in numerous countries and environmental conditions, thus providing a global perspective of the reproductive biology of this crop. It includes a bibliography based on papers from recent and old sources and unpublished perspectives from the authors.

This work is intended to become an important guide for undergraduate students but also an in-depth reference source for senior scientists. We have included numerous close-up photographs of various plant organs such as leaves, petioles, flowers, husks, fruits and pulp, seeds, and the plant as a whole, so that the reader is able to really "see" what the plant looks like (Figs. 2, 3, 4, 5, 6, 7 and 8). Large-format images are utilized to maximize the details of plant structures. These close-up photographs give a clear concept of the morphology of uchuva.

We hope readers will find the book interesting and pleasant to read. Every detail has been carefully reviewed to provide a cutting-edge understanding of the latest information.

Bogotá, Colombia   Fernando Ramírez

Homestead, FL, USA   Thomas Lee Davenport
October 2020

Fig. 1 Colombian uchuva ecotype. Photo taken at an orchard near Tuta, Boyacá State, Colombia

Fig. 2 Fruits covered by husk. Photo taken at Tuta, Boyacá State, Colombia

Fig. 3 Uchuva flower with dehiscent anthers. Photo take in Bogotá, Cundinamarca State, Colombia

Fig. 5 Stingless bee visiting uchuva flower in Bogotá, Cundinamarca State, Colombia

Fig. 4 An uchuva flower with purplish style, greenish stigma, and gray stamens. Photo taken in Bogotá, Cundinamarca State, Colombia

Fig. 6 Pubescences on uchuva stem in Bogotá, Cundinamarca State, Colombia

Fig. 7 Uchuva flower view from top taken in Bogotá, Cundinamarca State, Colombia. Note the pubescent calyx

Fig. 8 Uchuva flower with leaves and buds. Photo taken in in Bogotá, Cundinamarca State, Colombia

Acknowledgment

L. Marien for her kind support. Ms. Melanie van Overbeek of Springer for her excellent editorial advice and for making this book possible.

All photographs were taken by Fernando Ramírez unless stated otherwise. Photographs have been reproduced with permission.

Contents

Chapter 1
Introduction

Abstract Uchuva (*Physalis peruviana* L.) is an Andean plant that has attracted interest from numerous researchers worldwide. It belongs to the Solanaceae family. *P. peruviana* has been cultivated in tropical and subtropical environments in numerous countries worldwide, and it is considered an economically important plant. Environmental conditions are key for growth and development. Insects are important in pollination and fruit set. Self- and cross-pollination occur in uchuva. Germplasm repositories have been established and are essential for breeding and hybridization programs. The genetic diversity of uchuva has been analyzed using Random Amplified Microsatellites (RAMs) and Simple Sequence Repeats (SSR) among other molecular techniques. The morphological diversity of uchuva has been analyzed using quantitative and qualitative morphological features displayed in plants originating from different regions. It is considered an ever-bearing plant in the tropics that produces flowers and fruits constantly throughout the year. The phyllotaxy of the main leaves inserted in a stem is always alternate. Flowers are solitary and pendulous. Fruit development is a complex process that is influenced by temperature. Fruits are delicious and have been used in a number of recipes i.e. salads, sauces, preserves, jams, and jellies. This book covers various aspects of reproductive biology, including flowering, flower morphology, pollen morphology, pollination, self- and cross-pollination, floral visitation by insects and arthropods. It also includes seed viability, propagation, germination, plant development, fruit development and health benefits, breeding and hybridization, genetic diversity, chromosome number, and morphological diversity.

Physalis peruviana, a delicious uniquely flavored tropical fruit, has a number of common names that are culture-based and country or region specific. For example, Spanish names include uchuva (Colombia), popa o chimbonba (Nicaragua), topo-topo (Venezuela), bolsa mullaca (Peru), and many more according to regions and dialects (Lim 2013). The name, cape gooseberry, was adopted by Australians who obtained the plant from The Cape of Good Hope, South Africa (Morton 1987). Other English names include golden berry, angular winter cherry, balloon cherry, Peruvian

F. Ramírez, T. L. Davenport, *Uchuva (Physalis peruviana L.) Reproductive Biology*, https://doi.org/10.1007/978-3-030-66552-4_1

cherry, husked tomato, and ground cherry among many others. Uchuva is called joá-de-capote, balãozinho, bucho de ra, and camapú in Brazilian Portuguese. French names include batoto, alkékenge du Mexique, caqueret, coqueret anguleux, herbe á cloques, coqueret du Mexique, and more according to French-speaking region (Lim 2013). The plant is not considered a true gooseberry, and thus, we have used the name "uchuva" to refer to *Physalis peruviana*. The term, "uchuva", was coined by native indigenous people from Boyacá and Cundinamarca States, Colombia. We feel uchuva gives the plant a unique recognition and name derived from the Andes.

Uchuva evolved in the highlands of the Andes of South America (Fischer et al. 2011; Duarte and Paull 2015). The cultivation history of this plant can be traced back to the Incas (Klinac 1986). Uchuva was cultivated in Peru in the same areas where tomato originated (Legge 1974). Bartholomäus et al. (1990) claims its origin can be traced back to Ecuador and Peru, but Foqué (1973) extends its origin to the Andes of Venezuela, Colombia, and Chile between 800 and 3.000 m altitude. It is considered by some to be native to Peru and Chile (Morton 1987), but uchuva is indigenous to tropical high altitudes of Colombia, Chile, Ecuador, and Peru (Lim 2013).

Uchuva belongs to the Solanaceae, or nightshade family, which includes economically important tropical crops, such as lulo (*Solanum quitoense* Lam.), tree tomato (*Solanum betaceum* Cav.), and cocona (*Solanum sessiliflorum* Dunal) (Ramírez et al. 2018; Ramírez and Kallarackal 2019; Ramírez and Davenport 2020; Ramírez 2020, 2021), as well as crops that have spread to the subtropics, such as tomato (*Solanum lycopersicum* L.), eggplant (*Solanum melongena* L.), and potatoes (*Solanum tuberosum* L.) (Gil et al. 2019; Knapp et al. 2019; Gitari et al. 2018). The genus, *Physalis*, is characterized by an inflated fruiting calyx, which encloses the berry (Fig. 1.1) (Ganapathi et al. 1991; Ramírez et al. 2013). *Physalis* contains about 90 species, most of which occur in Mexico, the center of origin of this genus (Whitson and Manos 2005), but more recently, Duarte and Paull (2015) reported that *Physalis* has around 100 species. Historically, the genus *Physalis* has been separated into species groups on the basis of character i.e. growth habit, trichome type, and number of calyx angles (Rydberg 1896; Martínez 1999; Whitson and Manos 2005). *Physalis* has only a few representative species in the Old World, and a dozen species are native to South America (Brücher 1989).

Uchuva is an economically important horticultural plant that produces a delicious ornamental fruit of unique flavor and texture (Ramadan 2011). During 2018, Colombia exported uchuvas to countries such as the Netherlands (58%), Germany (11%), United Kingdom (11%), United States (8%), Canada (4%), Brazil (4%), and other countries (4%) (Procolombia 2018). Cultivated area has increased significantly during recent years in Colombia (Salazar et al. 2008) resulting in increased uchuva production from 11.1 tonnes in 2012 to 15.9 tonnes in 2017 (Procolombia 2018). Production also occurs in Ecuador, Kenya, Zimbabwe, New Zealand, India, Malaysia, China, USA (Hawaii and California), South Africa, Australia, and United Kingdom (National Research Council 1989; Vaughan and Geissler 2009; Fischer and Miranda 2012).

Uchuva is a herbaceous or soft-wooded, perennial plant reaching heights of one to 1.6 m and even two meters with sympodial growth (Fischer 2000; Morton 1987; Duarte and Paull 2015; Fischer and Almanza-Merchán, 2012; Fischer and Miranda

2012; Lim 2013). The plant is branched, with densely pubescent internodes, sometimes with purplish longitudinal lines or background color (Figs. 1.2, 1.3, 1.4, and 1.5). Leaves are velvety and heart-shaped with smooth edges and acuminate tip when young and become randomly toothed edges from 6 to 15 cm long and 4 to 10 cm wide when mature (Figs. 1.6, 1.7, 1.8, 1.9 and 1.10) (Morton 1987; Brücher

Fig. 1.1 Inflated fruiting calyces of uchuva. Top photo- immature calyx and bottom photo- mature calyx. Photos taken in Bogotá, Cundinamarca State, Colombia

Fig. 1.2 Branching pattern and pubescences on the stems of an adult uchuva plant. Photo taken in Bogotá, Cundinamarca State, Colombia

Fig. 1.3 Purplish mature stem with pubescences. Photo taken in Bogotá, Cundinamarca State, Colombia

Fig. 1.4 Stem pubescences. Top photo, underside of stem and bottom image, upper side of stem with purple color. Photos taken in Bogotá, Cundinamarca State, Colombia

Fig. 1.5 Pubescences in adult and young plants. Top image, upper side of stem with purplish lines in adult plant and bottom photo, gray pubescences in young plant. Photos taken in Bogotá, Cundinamarca State, Colombia

Fig. 1.6 Leaf morphology. Mature adaxial heart-shaped leaf. Photo taken in Bogotá, Cundinamarca State, Colombia

Fig. 1.7 Detail of leaf pubescences. Photo taken in Bogotá, Cundinamarca State, Colombia

Fig. 1.8 Adult leaf morphology. Abaxial leaf surface. Photo taken in Bogotá, Cundinamarca State, Colombia

Fig. 1.9 Close-up view of leaf petiole. Photo taken in Bogotá, Cundinamarca State, Colombia

Fig. 1.10 Young leaf morphology. Top photo, immature adaxial leaf surface. Lower photo, detail of pubescences. Photos taken in Bogotá, Cundinamarca State, Colombia

1989; Lim 2013; Duarte and Paull 2015). They have an alternate phyllotaxy (Figs. 1.11 and 1.12) (Fischer and Almanza-Merchán 2012). The plant's fibrous root system is found between 10 and 15 cm deep while some main roots reach 50–80 cm in depth (Duarte and Paull 2015).

Fig. 1.11 Leaf phyllotaxy in the 'Colombia' accession of uchuva showing alternating leaf insertion on adult stem. Top image is viewed from below. Lower image is side view. Photos taken in Bogotá, Cundinamarca State, Colombia

Fig. 1.12 Leaf phyllotaxy. Top image, alternate leaf petiole insertion into stem of an adult plant. Lower image, side view show alternate phyllotaxy of leaves in a young plant. Photos taken in Bogotá, Cundinamarca State, Colombia

The apical meristems are active throughout the plant's life. Vegetative and reproductive growth occurs simultaneously, thus producing leaves, new shoots, flowers, and fruit constantly. This plant is considered ever bearing and does not have resting periods after flower and fruiting begins.

Uchuva fruits are succulent, orange spheres with a pleasing taste that is acidic and sweet (National Research Council 1989). They are each protected by an inflated fruiting calyx or papery husk resembling a Chinese lantern. The attractive, symmetrical calyx with its edible yellow fruit inside gives it an eye-catching appearance and potential market appeal (National Research Council 1989). The fruit is globose, 1.25–2 cm wide, with glossy, smooth, orange-yellow skin, and juicy pulp containing numerous small seeds (Morton 1987). Fruits are usually eaten fresh out of hand (National Research Council, 1989). Chilled fruits provide a crisp, tartly sweet addition to both vegetable and fruit salads. Sometimes they are sweetened by pricking the skin and rolling them in sugar, which they absorb. Uchuvas are also used in glazes, sauces, preserves, jams and jellies, puddings, chutneys, ice cream, and fruit cocktails (Morton 1987; National Research Council 1989).

Environmental conditions regulate growth and development of uchuva plants. A minimum of 800 mm of rainfall is required for optimum growth (Duarte and Paull, 2015). Higher rainfall, about 4300 mm, has been found to increase growth and yield in well-drained soils. Precipitation required in commercial orchards of Colombia range from 800 to 1000 mm annually (Fischer and Almanza-Merchán 1993). Best plant production occurs with 1000–2000 mm of annual rainfall (Fischer et al. 2011).

Uchuva plants grow well when subjected to an annual average temperature of 13 °C–18 °C (Fischer 2000; Duarte and Paull 2015). It is adapted to temperatures ranging from 13 to 17 °C in the Andes (Fischer and Almanza-Merchán 1993). High daily temperatures of 27–30 °C do not reduce fruit set, but temperatures > 30 °C inhibit flowering and fruiting (Wolff 1991; Fischer 2000). Low temperatures are detrimental for plant growth and development. Freezing temperatures cause severe damage to uchuva plants (Fischer and Miranda 2012). They do not thrive at temperatures ≤10 °C (National Research Council 1989). Soil temperatures ≤8 °C limit growth and production, whereas higher temperatures, such as 22 °C, increase growth and fruit production (Fischer and Miranda 2012). Regarding relative humidity, uchuva plants require 70% to 80% (Fischer 2000).

Uchuva plants grow wild from 1500 to 3000 m above sea level (m a.s.l.) in Chile and Colombia (Duarte and Paull 2015). They are found in forests above 2200 m a.s.l. in Colombia (Fischer 1995). Preferred cultivated plant growth conditions in Colombia range between 1800 and 2800 m a.s.l. (Fischer 2000). They are smaller in height and have small, thick leaves at the highest altitudes. Cultivated plants grow more robustly and have better health conditions at 2400 m a.s.l. in Paipa and at 2700 m a.s.l. in the Tunja area of Boyacá State, Colombia (Fischer and Almanza-Merchán 1993). They are produced from 1500 to 3000 m a.s.l in most of the Andean countries (Fischer and Miranda 2012). Uchuva grows wild from 800 m a.s.l. in the coastal range to 3000 m a.s.l. in the Andes of Venezuela (Morton 1987) and from 300 a.s.l. to 2400 m a.s.l. in Hawaii and Ecuador (Duarte and Paull 2015). It is cultivated at altitudes ranging from 450 to 2030 m a.s.l. in India (Bala and Gupta

2011). In northern India, it is not possible to cultivate it above 1200 m a.s.l. due to the close proximity of the Himalayan mountains, but it thrives up to 1800 m a.s.l. in Southern India.

Uchuva plants grow in the full sun and also grow under shaded conditions (Duarte and Paull 2015). They perform better under full sunlight conditions in the tropics (Fischer and Miranda 2012). Influence of photoperiod on floral induction is equivocal. Although Heinze and Midash (1991) suggested uchuva is a quantitative, short-day plant, there is no substantial evidence to support this conclusion because the authors failed to investigate the impacts of short and long photoperiods on floral induction.

References

Bala S, Gupta RC (2011) Effect of secondary associations on meiosis, pollen fertility and pollen size in cape gooseberry (*Physalis peruviana* L.). Chromosom Bot 6:25–28. https://doi.org/10.1021/la2036746

Bartholomäus A, de la Rosa A, Santos J et al (1990) El manto de la tierra – Flora de los Andes. Lerner, Bogotá

Brücher H (1989) Useful plants of Neotropical origin and their wild relatives. Springer, Mendoza

Duarte O, Paull R (2015) Exotic fruits and nuts of the new world. CABI, Wallingford

Fischer G (1995) Effect of root zone temperature and tropical altitude on the growth, development and fruit quality of cape gooseberry (*Physalis peruviana* L.). PhD Tesis. Humboldt-Universität zu Berlin, Berlin

Fischer G (2000) Crecimiento y desarrollo. In: Flórez V, Fischer G, Sora A (eds) Producción, poscosecha y exportación de la uchuva (*Physalis peruviana* L.). Universidad Nacional de Colombia, Unibiblos, Bogotá, pp 9–26

Fischer G, Almanza-Merchán P (1993) La uchuva (*Physalis peruviana* L.) una alternativa promisoria para las zonas altas de Colombia. Agric Trop 30:79–87

Fischer G, Almanza-Merchán P (2012) Fisiología del cultivo de la uchuva (*Physalis peruviana* L.). In: II Reuniao Técnica da Cultura da Physalis. Lages, pp 32–52

Fischer G, Miranda D (2012) Uchuva (*Physalis peruviana* L.). In: Fischer G (ed) Manual para el cultivo de frutales en el trópico. Produmedios, Bogotá, pp 851–873

Fischer G, Herrera A, Almanza PJ (2011) Cape gooseberry (*Physalis peruviana* L.). In: Yahia EM (ed) Postharvest biology and technology of tropical and subtropical fruits. Woodhead Publishing, Oxford, pp 374–397

Foqué A (1973) Espèces fruitières d'Amérique tropicale. Fruits 28:40–49

Ganapathi A, Sudhakaran S, Kulothungan S (1991) The diploid taxon in indian natural populations of *Physalis* L. and its taxonomic significance. Cytologia (Tokyo) 56:283–288

Gil R, Bojacá CR, Schrevens E (2019) Understanding the heterogeneity of smallholder production systems in the Andean tropics – the case of Colombian tomato growers. NJAS – Wagening J Life Sci 88:1–9. https://doi.org/10.1016/J.NJAS.2019.02.002

Gitari HI, Gachene CKK, Karanja NN et al (2018) Optimizing yield and economic returns of rainfed potato (*Solanum tuberosum* L.) through water conservation under potato-legume intercropping systems. Agric Water Manag 208:59–66. https://doi.org/10.1016/J.AGWAT.2018.06.005

Heinze W, Midash M (1991) Photoperiodische Reaktion von *Physalis peruviana* L. Gartenbauwissenschaft 56:262–264

Klinac DJ (1986) Cape gooseberry (*Physalis peruviana*) production systems. N Z J Exp Agric 14:425–430. https://doi.org/10.1080/03015521.1986.10423060

Knapp S, Aubriot X, Prohens J (2019) Eggplant (*Solanum melongena* L.): taxonomy and relationships. In: Chapman M (ed) The eggplant genome. Springer, Cham, pp 11–22

Legge A (1974) Notes on the history, cultivation and uses of *Physalis peruviana* L. J R Hortic Soc 99:310–314

Lim TK (2013) Edible medicinal and non-medicinal plants. Springer, Dordrecht

Martínez M (1999) Infrageneric taxonomy of *Physalis*. In: Nee M, Symon D, Lester R, Jessop J (eds) Solanaceae IV: advances in biology and utilization. Royal Botanic Gardens, Kew, pp 275–283

Morton J (1987) Cape gooseberry. In: Fruits of warm climates. Echo Point Books & Media, Miami, pp 430–434

National Research Council (1989) Lost crops of the Incas: little-known plants of the Andes with promise for worldwide cultivation. National Academy Press, Washington, DC

Procolombia (2018) Uchuva (goldenberry). https://docs.procolombia.co/int-procolombia/es/exportaciones/ficha_uchuva_final.pdf

Ramadan MF (2011) Bioactive phytochemicals, nutritional value, and functional properties of cape gooseberry (*Physalis peruviana*): an overview. Food Res Int 44:1830–1836

Ramírez F (2020) Cocona (*Solanum sessiliflorum* Dunal) reproductive physiology: a review. Genet Resour Crop Evol 67:293–311. https://doi.org/10.1007/s10722-019-00870-x

Ramírez F (2021) Notes about Lulo (*Solanum quitoense* Lam.): an important South American underutilized plant. Genet Resour Crop Evol 68:93–100. https://doi.org/10.1007/s10722-020-01059-3

Ramírez F, Davenport TL (2020) The development of lulo plants (*Solanum quitoense* Lam. var. *septentrionale*) characterized by BBCH and landmark phenological scales. Int J Fruit Sci 20:562–585. https://doi.org/10.1080/15538362.2019.1613470

Ramírez F, Kallarackal J (2019) Tree tomato (*Solanum betaceum* Cav.) reproductive physiology: a review. Sci Hortic (Amsterdam) 248:206–215

Ramírez F, Fischer G, Davenport TL et al (2013) Cape gooseberry (*Physalis peruviana* L.) phenology according to the BBCH phenological scale. Sci Hortic (Amsterdam) 162:39–42. https://doi.org/10.1016/j.scienta.2013.07.033

Ramírez F, Kallarackal J, Davenport TL (2018) Lulo (*Solanum quitoense* Lam.) reproductive physiology: a review. Sci Hortic (Amsterdam) 238:163–176. https://doi.org/10.1016/j.scienta.2018.04.046

Rydberg P (1896) The North American species of *Physalis* and related genera. Mem Torrey Bot Club 4:297–372

Salazar MR, Jones JW, Chaves B, Cooman A (2008) A model for the potential production and dry matter distribution of cape gooseberry (*Physalis peruviana* L.). Sci Hortic (Amsterdam) 115:142–148. https://doi.org/10.1016/j.scienta.2007.08.015

Vaughan J, Geissler C (2009) New Oxford fook of food plants. Oxford University Press, Oxford

Whitson M, Manos PS (2005) Untangling *Physalis* (Solanaceae) from the physaloids: a two-gene phylogeny of the physalinae. Syst Bot 30:216–230. https://doi.org/10.1600/0363644053661841

Wolff XY (1991) Species, cultivar, and soil amendments influence fruit production of two *Physalis* species. HortScience 26:1558–1559

Chapter 2
Flower Morphology

Abstract Uchuva flowers are campanulate with fused petals that form five distinct lobes. The corolla is yellow with five prominent brown spots and trichomes on the outer surface. The five sepals are green with copious white trichomes outside and purple venation on the inner side. The style is claviform, and the stigma is capitate and light green. The five anthers of young uchuva flowers are dark purplish-blue and oblong-shaped, whereas mature anthers are brown. They bear two longitudinal slits on their surface that split or dehisce to release pollen. Following anther dehiscence the entire corolla, including petals and stamens, senesces.

Newly emerging uchuva floral buds are ovoid, brown, and pubescent (Figs. 2.1 and 2.2) (Fischer 2000; Ligarreto et al. 2005). Flowers are campanulate wherein the petals are fused to form five distinct lobes in the shape of a flared bell (Figs. 2.3 and 2.4) (Ramírez et al. 2013; Duarte and Paull 2015). The uchuva flower is similar to other *Solanum* species (Ramírez et al. 2018; Ramírez and Kallarackal 2019; Ramírez and Davenport 2020; Ramírez 2020, 2021). The background color of the corolla is yellow with five prominent brown spots or maculations linked to brown veins along the center of each lobe (Figs. 2.3, 2.4, 2.5, 2.6, 2.7, 2.8, 2.9, 2.10, 2.11 and 2.12) (Lagos et al. 2005a, b, 2008; Bhat et al. 2018). The outer surface of the corolla bears small white trichomes whereas the inner surface is glabrous (Figs. 2.3, 2.4, 2.5 and 2.6) (Lagos et al. 2005a, b, 2008). Corolla size ranges from 1.2 to 1.8 cm in diameter, and the brown maculations are about a quarter of that (Figs. 2.3 and 2.7) (Lagos et al. 2005a, b, 2008). The five campanulate sepals are green bearing copious white trichomes outside, purple venation on the inner side, and will enlarge to form the calyx or husk that will ultimately cover the developing fruit (Figs. 2.5 and 2.6) (He and Saedler 2005; Lagos et al. 2008; Fischer et al. 2014). Although not visible in the currently presented figures, the ovary is superior, yellow-green or green (Lagos et al. 2008). It is 2.5 mm in length and 2.6 mm in diameter (Lagos et al. 2005b). The single style is claviform (club-shaped) and deep purple in color, and the stigma is capitate and light green (Figs. 2.7, 2.8, 2.9, 2.10 and 2.11) (Ligarreto et al. 2005).

© The Author(s), under exclusive license to Springer Nature
Switzerland AG 2021
F. Ramírez, T. L. Davenport, *Uchuva (Physalis peruviana L.) Reproductive Biology*, https://doi.org/10.1007/978-3-030-66552-4_2

Fig. 2.1 Floral buds. Upper photo, side view and lower photo, top view. Note conspicuous pubescences along the bud and petiole. Photos taken in Bogotá, Cundinamarca State, Colombia

Fig. 2.2 Floral buds and leaves in young plant. Upper photo, side view and lower photo, top view. Note prominent pubescences on leaves and buds. Photos taken in Bogotá, Cundinamarca State, Colombia

Fig. 2.3 Fused petals and blue anthers in young flower. Photo taken in Bogotá, Cundinamarca State, Colombia

Fig. 2.4 Mature uchuva flower. Underside of flower with stamens, yellow-fused petals with brown maculations. Photo taken in Bogotá, Cundinamarca State, Colombia

Fig. 2.5 Lateral view of pendulous uchuva flower. Note the pubescent peduncle. Photo taken in Bogotá, Cundinamarca State, Colombia

Fig. 2.6 Uchuva sepals, petals, and maculations viewed from above. Note the white trichomes on the outside surface of the sepals. Photo taken in Bogotá, Cundinamarca State, Colombia

Fig. 2.7 Young uchuva flower with yellow fused petals (1), brown maculations (2), green-colored stigma on top of dark purple style (3), and surrounded by blue-purple anthers (4), and greenish sepal (5). Photo taken in Bogotá, Cundinamarca State, Colombia

Fig. 2.8 Front view of uchuva flower. Purplish style (1), greenish stigma (2), dehisced anther reveling pollen (3). Photo taken in Bogotá, Cundinamarca State, Colombia

Fig. 2.9 Uchuva flower. Note the greenish stigma between the two stamens and fine hairs surrounding the stamens at their base. Photo taken in Bogotá, Cundinamarca State, Colombia

Fig. 2.10 Flower with purple style and greenish stigma. Photo taken in Bogotá, Cundinamarca State, Colombia

Fig. 2.11 Side view of post-dehiscent flower. Note the position of the style, stigma, and stamens. Photo taken in Bogotá, Cundinamarca State, Colombia

Fig. 2.12 Anther dehiscence. Note the copious white pollen grains on the anther and lower maculation surfaces. Also note the distinct purple venations extending from each maculation to petal tip demarking the center of each fused petal or lobe. Photo taken in Bogotá, Cundinamarca State, Colombia

The sepals are 9.5 mm–12 mm in length and 6 mm in diameter (Fig. 2.6) (Benitez and Magallanes 1998; Ligarreto et al. 2005; Bhat et al. 2018). Young uchuva flowers bear five dark purplish-blue, oblong-shaped anthers, whereas mature anthers are brown (Figs. 2.3, 2.4, 2.5, and 2.7) (Ligarreto et al. 2005; Fernandes da Silva et al. 2017). The anthers are 4 mm long and 3–4 mm wide (Ligarreto et al. 2005; Lagos et al. 2008; Perea et al. 2010; Bhat et al. 2018). They have two longitudinal slits on their surface that split or dehisce to release pollen (Figs. 2.10, 2.11, 2.12 and 2.13). The average number of pollen grains per anther is about 1900, and the mean number of pollen grains per flower is over 9600 (Fernandes da Silva et al. 2017).

Individual flowers develop at the base of each leaf petiole (Fischer et al. 2011; Barboza et al. 2016); and bear pedicels that bend geotropically downward so as to hang pendulously as they elongate reaching up to 0.7–1.1 cm in length (Ganapathi et al. 1991; Ligarreto et al. 2005; Bhat et al. 2018) (Figs. 2.5 and 2.6). Perea et al. (2010) reported extremely long floral pedicels up to 3–5 cm in length, which seem to be outside the norm for uchuvas grown in Colombia.

Following anther dehiscence and shedding of the pollen (Figs. 2.12 and 2.13), the entire corolla, including petals and stamens, begins to senesce and abscise as a unit from the receptacle at the base of the ovary (Figs. 2.14, 2.15, 2.16, 2.17 and 2.18). Separation of the corolla leaves exposed the developing four-sector ovary fruit with residual style (Figs. 2.19, 2.20 and 2.21). The latter soon abscises (Fig. 2.22) as the elongating calyx envelops the growing fruit. The surrounding calyx continues to enlarge forming a protective envelope for the developing fruit as it grows (Fig. 2.23).

Fig. 2.13 Anther dehiscence. Photo taken in Bogotá, Cundinamarca State, Colombia

Fig. 2.14 Corolla senescence. Note how the corolla is about to fall. Photo taken in Chipaque, Cundinamarca State, Colombia

Fig. 2.15   Fallen corolla and stigma. Photo taken in Bogotá, Cundinamarca State, Colombia

Fig. 2.16 Abscised petals side view. Photo taken in Bogotá, Cundinamarca State, Colombia

Fig. 2.17 Abscised petals. Note base of the desiccated corolla. Photo taken in Bogotá, Cundinamarca State, Colombia

Fig. 2.18 Abscised corolla cut to reveal the discolored anthers. Photo taken in Bogotá, Cundinamarca State, Colombia

Fig. 2.19 Residual style protruding from calyx. Photo taken in Bogotá, Cundinamarca State, Colombia

Fig. 2.20 Calyx with residual desiccated style. Photo taken in Bogotá, Cundinamarca State, Colombia

Fig. 2.21 Residual style attached to ovary. Photo taken in Bogotá, Cundinamarca State, Colombia

Fig. 2.22 Four sector ovary after style abscission and before calyx envelopment of the fruit. Photo taken in Bogotá, Cundinamarca State, Colombia

Fig. 2.23 Calyx envelops the growing fruit, ultimately sealing the fruit inside. Photo taken in Bogotá, Cundinamarca State, Colombia

References

Barboza G, Hunziker A, Bernardello G et al (2016) Solanaceae. In: Kadereit J, Bittrich V (eds) Flowering plants, Eudicots, the families and genera of vascular plants 14. Springer, Cham, pp 295–358

Benitez C, Magallanes A (1998) El género *Physalis* (Solanaceae) de Venezuela. Acta Bot Venezuélica 21:11–42

Bhat N, Jeri L, Mipun P, Kumar Y (2018) Systematic studies (micro-morphological, leaf architectural, anatomical and palynological) of genus *Physalis* L. (Solanaceae) in Northeast India. Plant Arch 18:2229–2238

Duarte O, Paull R (2015) Exotic fruits and nuts of the new world. CABI, Wallingford

Fernandes da Silva D, Pio R, Vieira Nogueira P et al (2017) Viabilidade polínica e quantificação de grãos de pólen em espécies de fisális. Rev Ciênc Agron 48:365–373

Fischer G (2000) Crecimiento y desarrollo. In: Flórez V, Fischer G, Sora A (eds) Producción, poscosecha y exportación de la uchuva (*Physalis peruviana* L.). Universidad Nacional de Colombia, Unibiblos, Bogotá, pp 9–26

Fischer G, Herrera A, Almanza PJ (2011) Cape gooseberry (*Physalis peruviana* L.). In: Yahia EM (ed) Postharvest biology and technology of tropical and subtropical fruits. Woodhead Publishing, Oxford, pp 374–397

Fischer G, Almanza-Merchán PJ, Miranda D (2014) Importancia y cultivo de la uchuva (*Physalis peruviana* L.). Rev Bras Frutic 36:001–015. https://doi.org/10.1590/0100-2945-441/13

Ganapathi A, Sudhakaran S, Kulothungan S (1991) The diploid taxon in indian natural populations of *Physalis* L. and its taxonomic significance. Cytologia (Tokyo) 56:283–288

He C, Saedler H (2005) Heterotopic expression of MPF2 is the key to the evolution of the Chinese lantern of *Physalis*, a morphological novelty in Solanaceae. Proc Natl Acad Sci U S A 102:5779. https://doi.org/10.1073/pnas.0501877102

Lagos T, Caetano C, Vallejo F et al (2005a) Caracterización palinológica y viabilidad polínica de *Physalis peruviana* L. y *Physalis philadelphica* Lam. Agron Colomb 23:55–61. ISSN 0120-9965

Lagos T, Criollo H, Paredes O et al (2005b) Estudio de la biologia floral de la uchuva (*Physalis peruviana* L.). Rev Cienc Agric 22:1–11

Lagos T, Vallejo Cabrera A, Escobar C, Muñoz Flórez J (2008) Biología reproductiva de la uchuva. Acta Agron 57:81–87

Ligarreto G, Lobo M, Correa M (2005) Recursos genéticos del género *Physalis* en Colombia. In: Fischer G, Miranda D, Piedrahita W, Romero J (eds) Avances en cultivo, poscosecha y exportación de la uchuva (*Physalis peruviana* L.) en Colombia. Universidad Nacional de Colombia, Unibiblos, Bogotá, pp 9–26

Perea M, Rodríguez N, Fischer G et al (2010) Uchuva: *Physalis peruviana* L. (Solanaceae). In: Perea M, Matallana L, Tirado A (eds) Biotecnología aplicada al mejoramiento de los cultivos de frutas tropicales. Universidad Nacional de Colombia, Bogotá, pp 466–490

Ramírez F (2020) Cocona (*Solanum sessiliflorum* Dunal) reproductive physiology: a review. Genet Resour Crop Evol 67:293–311

Ramírez F (2021) Notes about Lulo (*Solanum quitoense* Lam.): an important South American underutilized plant. Genet Resour Crop Evol 68:93–100. https://doi.org/10.1007/s10722-020-01059-3

Ramírez F, Davenport TL (2020) The development of lulo plants (*Solanum quitoense* Lam. var. *septentrionale*) characterized by BBCH and landmark phenological scales. Int J Fruit Sci 20:562–585. https://doi.org/10.1080/15538362.2019.1613470

Ramírez F, Kallarackal J (2019) Tree tomato (*Solanum betaceum* Cav.) reproductive physiology: a review. Sci Hortic (Amsterdam) 248:206–215

Ramírez F, Fischer G, Davenport TL et al (2013) Cape gooseberry (*Physalis peruviana* L.) phenology according to the BBCH phenological scale. Sci Hortic (Amsterdam) 162:39–42. https://doi.org/10.1016/j.scienta.2013.07.033

Ramírez F, Kallarackal J, Davenport TL (2018) Lulo (*Solanum quitoense* Lam.) reproductive physiology: a review. Sci Hortic (Amsterdam) 238:163–176. https://doi.org/10.1016/j.scienta.2018.04.046

Chapter 3
Phyllotaxy and Floral Development

Abstract During germination, the curved hypocotyl pushes up though the soil, then the seedling plant elongates linearly upward. It then transitions to a form of sympodial growth that results in a bifurcating branch pattern. This represents the transition from a juvenile to mature plant. Each uchuva stem has a primary leaf, flower and two lateral meristems resulting in frequent branching. The phyllotaxy of the main leaves and spurs inserted in a stem is always alternate. Floral, vegetative, and fruit development thus occur simultaneously. The plant growth and development patterns of uchuva are characterized by a phenological scale.

In nature, every uchuva plant begins with germination of a viable seed (See Chap. 7 or details on germination). This seedling plant elongates linearly upward behind a single apical meristem forming a leaf at each node with no side branches (Fig. 3.1). This juvenile growth pattern continues until the plant is about 20 cm in length (Fig. 3.2). It then transitions to a form of sympodial growth that results in a bifurcating branch pattern wherein one branch develops more strongly than the other to further extend the primary stem, and the weaker branch develops into a lateral spur (Fig. 3.3) (von Sachs 1882).

Bifurcation of the main stem is the start point for subsequent sympodial growth and flowering. It represents the transition from a juvenile to mature plant (Ramírez et al. 2013). Uchuva stem elongation is a process that involves development in the apical meristem of a node by formation of a primary, main-stem leaf with its supporting petiole. At its base is a floral meristem, forming a single flower, and two lateral meristems that form bifurcating shoots at the petiole base (Figs. 3.4, 3.5, 3.6 and 3.7). One stem is dominant to continue the main stem at a slight angle to the previous internode, and the other forming a lateral spur that projects at a 45° angle to the main-stem leaf petiole (Figs. 3.8 and 3.9). This process repeats itself with extension of each stem internode contributing to further sympodial stem growth (Fig. 3.10). Thus, there is a flower at the base of each leaf at each node in a stem (Fischer 2000). Sometimes a spur will grow to form a lateral branch point. The phyllotaxy of the main leaves inserted in a stem is always alternate (Figs. 3.8 and 3.9).

F. Ramírez, T. L. Davenport, *Uchuva (Physalis peruviana L.) Reproductive Biology*, https://doi.org/10.1007/978-3-030-66552-4_3

45

Fig. 3.1 Elongating seedling of uchuva. Note the two expanded cotyledons. Photo taken in Bogotá, Cundinamarca State, Colombia

Fig. 3.2 Linear growth of young uchuva plant. Note that each true leaf is formed at nodes forming above the two lower most, opposing cotolydonary leaves. Photo taken in Bogotá, Cundinamarca State, Colombia

Fig. 3.3 Adult uchuva stem. Note that the left branch extending from the most proximal visible node (**a**) develops additional nodes (**b** and **c**) whereas the right branch (**d**) forms a terminal flowering spur. Photo taken in Bogotá, Cundinamarca State, Colombia

Fig. 3.4 Uchuva plant showing bifurcation. Photo taken in Bogotá, Cundinamarca State, Colombia

Fig. 3.5 Uchuva stem showing primary leaf, flower, and two lateral meristems. Photo taken in Bogotá, Cundinamarca State, Colombia

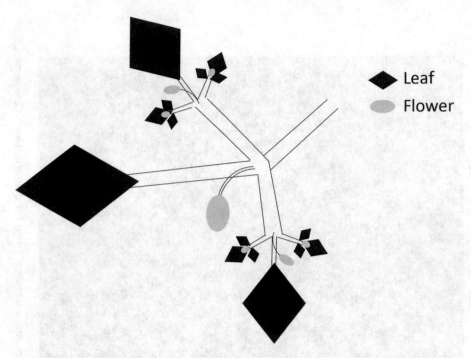

Fig. 3.6 Illustration showing stem with leaf, flower, and floral bud and two lateral meristems with primordial leaves and flowers

Flower development begins with the newly emerging single floral bud at the base of each leaf petiole (Fig. 3.11). Firstly, the floral bud increases in size and turns brownish green in color (Fig. 3.11). Then the sepals begin to open, and the petals become visible (Figs. 3.12 and 3.13). Next, the petals begin to elongate, and the floral corolla opens (Figs. 3.14, 3.15 and 3.16). Floral, vegetative, and fruit growth and development, thus, occur simultaneously (Ramírez et al. 2013).

The phenology of uchuva has been characterized using the BBCH (Biologische Bundesantalt, Bundessortenamt und Chemische Industrie) scale (Ramírez et al. 2013). This scale depicts two principal stages related to flowering; principal growth stage 5, called inflorescence emergence and stage 6, named flowering.

Tropical solanaceae species, including lulo, cocona and tree tomato, flower continuously through the year (Ramírez et al. 2018; Ramírez and Kallarackal 2019; Ramírez and Davenport 2020; Ramírez 2020, 2021). Similarly, uchuva flowering occurs throughout the year in warmer areas. The period from floral induction to anthesis has been recorded for uchuva growing in different countries (Table 3.1) (Gupta and Roy 1981; National Research Council 1989; Mazorra et al. 2003;

Fig. 3.7 Uchuva stem with open flower facing back, primary leaf, and two lateral meristems in center of image. Photo taken in Bogotá, Cundinamarca State, Colombia

Fig. 3.8 Alternate phyllotaxy. Upper photo, branch with intact leaves. Lower photo, leaves on spurs removed. Note the alternate disposition of main-stem leaves. Photos taken in Bogotá, Cundinamarca State, Colombia

Fig. 3.9 Upper photo, lateral view of alternate phyllotaxy on main stem. Lower photo, longitudinal view of main stem. Leaves on spurs were removed. Photos taken in Bogotá, Cundinamarca State, Colombia

Fig. 3.10 Uchuva plant with three stems. Photo taken in Bogotá, Cundinamarca State, Colombia

Fig. 3.11 Early floral buds. Upper photo, newly emerging floral bud. Lower photo, larger brown-green floral bud. Photos taken in Bogotá, Cundinamarca State, Colombia

Fig. 3.12 Upper photo, sepals start to separate and petals become visible. Lower photo, sepal separate to form five lobes. Photos taken in Bogotá, Cundinamarca State, Colombia

Fig. 3.13 Upper photo, petals are longer. Lower photo, view from below. Photos taken in Bogotá, Cundinamarca State, Colombia

Fig. 3.14 Petals begin to open and green stigma is visible. Photos taken in Bogotá, Cundinamarca State, Colombia

Fig. 3.15 Upper photo, side view. Lower photo, petals continue to open, stigma and anthers are visible. Photos taken in Bogotá, Cundinamarca State, Colombia

Fig. 3.16 Upper photo, petals open, and floral parts are visible. Lower photo, fully open flower. Photo taken in Bogotá, Cundinamarca State, Colombia

Table 3.1 Flowering in *Physalis peruviana*

Days after transplant to flowering	Country	References
70–80	India	Gupta and Roy (1981)
65–75 after sowing	Not specified	National Research Council (1989)
42	Chiapas, Mexico	Mora-Aguilar et al. (2006)
108	Brazil	Leitzke Betemps et al. (2014)
Period from floral initiation to anthesis		
3 weeks	Not specified	Duarte and Paull (2015)
19–23 days	India	Gupta and Roy (1981)
18–21 days	Sumapaz region, Colombia	Mazorra et al. (2003)
37 days	Nariño, Colombia	Lagos et al. (2008)

Mora-Aguilar et al. 2006; Lagos et al. 2008; Leitzke Betemps et al. 2014; Duarte and Paull 2015).

Flowers open in the morning and close in the evening of the same day (Gupta and Roy 1981). Flowers in Bogotá, Colombia begin opening at about 6:15 am and completely open by 9:30 am (Ramírez unpublished data). These same flowers closed at about 4–5 pm; however, some flowers have been observed to close after 6 pm. Lagos et al. (2008) examined floral opening times at a high-altitude tropical site located 2820 m a.s.l. in Nariño State, Colombia. They found that 98% of the flowers opened from 7:00 to 10:00 am, and 2% of the flowers in the same group of plants opened the following day at the same time.

Little is known about the dynamics of stigma receptivity or anther dehiscence to release pollen. Lagos et al. (2008) reported that stigmas are supposedly receptive 2 days before flower opening, thus limiting self-pollination but is unclear how they came to these conclusions. They also reported that anther dehiscence occurred soon after floral opening.

Flowering can be delayed or halted at high temperatures ≥ 30 °C (Fischer and Melgarejo 2014) whereas temperatures near or below 0 °C will cause necrosis on the adaxial leaf surface or abscission (Zapata et al. 2002).

Plants grow in full sun but can also grow in partial shade (Duarte and Paull 2015). They produce fruit near the equator and at high latitudes (National Research Council 1989). Some think uchuva is not photoperiodic and that this factor does not influence flowering (Duarte and Paull 2015). Mora-Aguilar et al. (2006), however, suggested that photoperiod in October–December could be linked to floral induction in Mexico. Furthermore, Heinze and Midasch (1991) reported that uchuva is a quantitative short-day plant.

Antúnez-Ocampo et al. (2016) examined the possible role of the ratio of ammonium and nitrate ions and days after pruning on uchuva flowering but they determined that the number of flowers produced in response to the various treatments was best correlated with plant vigor or rate of growth. This response is consistent with the observations described above that a single flower is always produced at the base of new primary leaves on main branches. High vigor, thus, produces more leaves and with them come the associated flowers.

References

Antúnez-Ocampo M, Sandoval-Villa M, Alcántar-González G et al (2016) Floración y fructificación de *Physalis peruviana* L. por la aplicación de amonio y nitrato, edad de vigor de la planta. Agrociencia 50:603–615

Duarte O, Paull R (2015) Exotic fruits and nuts of the new world. CABI, Wallingford

Fischer G (2000) Crecimiento y desarrollo. In: Flórez V, Fischer G, Sora A (eds) Producción, poscosecha y exportación de la uchuva (*Physalis peruviana* L.). Universidad Nacional de Colombia, Unibiblos, Bogotá, pp 9–26

Fischer G, Melgarejo L (2014) Ecofisiología de la uchuva (*Physalis peruviana* L.). In: Carvalho C (ed) Uchuva (*Physalis peruviana* L.) fruta Andina para el mundo. Limencop S.L, Alicante, pp 29–47

Gupta S, Roy S (1981) The floral biology of cape gooseberry (*Physalis peruviana* Linn; Solanaceae, India). Indian J Agric Sci 51:353–355

Heinze W, Midasch M (1991) Photoperiodic reaction of *Physalis peruviana*. Gartenbauwissenschaft 56:262–264

Lagos T, Vallejo Cabrera A, Escobar C, Muñoz Flórez J (2008) Biología reproductiva de la uchuva. Acta Agron 57:81–87

Leitzke Betemps D, Fachinello J, Madruga Lima C et al (2014) Época de semeadura, fenologia e crescimento de plantas de fisális no sul do Brasil. Rev Bras Frutic 36:179–185

Mazorra M, Quintana A, Miranda D et al (2003) Análisis sobre el desarrollo y la madurez fisiológica del fruto de la uchuva (*Physalis peruviana* L.) en la zona de Sumapaz (Cundinamarca). Agron Colomb 21:175–189

Mora-Aguilar R, Peña-Lomelí A, López-Gaytán E et al (2006) Agrofenología de *Physalis peruviana* L. en invernadero y fertirriego. Rev Chapingo Ser Hortic 12:57–63

National Research Council (1989) Lost crops of the Incas: little-known plants of the Andes with promise for worldwide cultivation. National Academy Press, Washington, DC

Ramírez F (2020) Cocona (*Solanum sessiliflorum* Dunal) reproductive physiology: a review. Genet Resour Crop Evol 67:293–311

Ramírez F (2021) Notes about Lulo (*Solanum quitoense* Lam.): an important South American underutilized plant. Genet Resour Crop Evol 68:93–100. https://doi.org/10.1007/s10722-020-01059-3

Ramírez F, Davenport TL (2020) The development of lulo plants (*Solanum quitoense* Lam. var. *septentrionale*) characterized by BBCH and landmark phenological scales. Int J Fruit Sci 20:562–585. https://doi.org/10.1080/15538362.2019.1613470

Ramírez F, Kallarackal J (2019) Tree tomato (*Solanum betaceum* Cav.) reproductive physiology: a review. Sci Hortic (Amsterdam) 248:206–215

Ramírez F, Fischer G, Davenport TL et al (2013) Cape gooseberry (*Physalis peruviana* L.) phenology according to the BBCH phenological scale. Sci Hortic (Amsterdam) 162:39–42. https://doi.org/10.1016/j.scienta.2013.07.033

Ramírez F, Kallarackal J, Davenport TL (2018) Lulo (*Solanum quitoense* Lam.) reproductive physiology: a review. Sci Hortic (Amsterdam) 238:163–176. https://doi.org/10.1016/j.scienta.2018.04.046

von Sachs J (1882) Placentation in angiosperms. Bot Rev 18:603–651

Zapata J, Saldarriaga A, Londoño M, Díaz C (2002) Manejo del cultivo de la uchuva en Colombia. Corpoica, Rionegro.

Chapter 4
Pollen Morphology

Abstract Uchuva pollen is a monad 17–33 μm in length, tricolporate, isopolar, and spheroidal in shape with three colporus apertures. The apertures in the pollen grains are arranged irregularly. Pollen viability in different uchuva accessions ranged 54.12–98.82% after 4 °C storage for 2 days. Pollen germination ranged from 3.70 to 53% when stored at 21 °C for 4 h. Pollen viability was highest (96%) in anthers collected soon after anthesis during the morning hours and decreased progressively through the day. Fresh pollen collected from flowers remained viable (97%) for 35–40 days when stored at 13 °C.

Solanaceous pollen grains vary in shape, apertures, size, aggregation, and exine ornamentation (Barboza et al. 2016; Ramírez et al. 2013, 2018; Ramírez and Kallarackal 2019; Ramírez and Davenport 2020; Ramírez 2020, 2021). Uchuva pollen is a monad 17–33 μm in length, tricolporate, isopolar, and spheroidal in shape with three colporus apertures (Table 4.1). Ganapathi et al. (1991) reported that pollen size ranges from 15 to 21 μm in uchuva plants growing in India. Benitez and Magallanes (1998) reported that pollen grains in flowers of uchuva plants growing in Venezuela are, on average, 17.7 μm in length and 11.8 μm in width. Pollen grains of some uchuva accessions, e.g. from Kenya are tetracolporate (Lagos et al. 2005). Both polar diameter and equatorial diameter ranged between 27 and 39 μm in Kenyan and UN-49 uchuva accessions (Table 4.1). The mean polar diameter of the Kenyan accession was 31 μm and the mean equatorial diameter was 32 μm (Lagos et al. 2005). In the Colombian accession, UN-49, the mean polar diameter was 33 μm and the mean equatorial diameter was 30 μm. da Cruz-Barros et al. (2011) reported that pollen grains observed in Brazil are prolate-spheroidal shaped with apertures of mid-constriction. The polar diameter is 34 μm, and the equatorial is diameter 32 μm. Pollen grains having irregularly arranged apertures have been frequently observed in Germany (Halbritter 2016). Small crystals have been observed attached to pollen and to the locular wall (Halbritter 2016). Bhat et al. (2018) described uchuva pollen from north-east India as a monad, oblate-spheroidal in shape, with three 4-zonocolporate apertures, radially symmetrical and with microechinate exine ornamentation (Fig. 4.1). The polar axis length was 23.7 ± 0.52 μm and equatorial axis width was 25.1 ± 0.96 μm. El-Ghamery et al. (2018) reported

F. Ramírez, T. L. Davenport, *Uchuva (Physalis peruviana L.) Reproductive Biology*, https://doi.org/10.1007/978-3-030-66552-4_4

Table 4.1 Pollen characteristics. Descriptions are from the various authors

Pollen unit	Shape	Aperture	Lenght (µm)	Width (µm)	Ecotype	References
			medium-size 15-21 µm			Ganapathi et al. (1991)
			17.7	11.8		Benitez and Magallanes (1998)
			33.0	30.0	UN-49	Lagos et al. (2005)
			31.5	31.6	Kenya	Lagos et al. (2005)
	Oblate-spheroidal		34	32		da Cruz-Barros et al. (2011)
Monad	Spheroidal	Colporate, tricolporate	medium-size 26-50 µm			Halbritter (2016)
Monad	Oblate-spheroidal		23.7	25.1		Bhat et al. (2018)
			25–25.9	24.4–25.6		El-Ghamery et al. (2018)
		Tricolporate	18–24	18–20		Finot et al. (2018)
Monad	Spheroidal		27.5	26.3		Fernandes da Silva et al. (2019)

that the polar diameters of uchuva pollen grains observed in Egypt ranged from 25 to 25.9 µm and equatorial diameters of 24.4–25.6 µm. Finot et al. (2018) found that uchuva pollen collected in Chile was 18–24 µm long and 18–20 µm wide. It is tricolporate, each colpi being 18–24 × 2–3 µm with tapering ends that are somewhat constricted equatorially. The apertural membrane is described as being densely covered by granular sexinous processes with lalongate apertures 2 × 9–14 µm and polar margins, usually constricted at the polar axis and with acute equatorial ends (Finot et al. 2018). Fernandes da Silva et al. (2019) reported that uchuva pollen is a monad 27.5 µm long and 26.3 µm wide when the plants were grown in direct sun light, but the pollen size and shape varied when the sun was blocked by shade cloths of different colors.

4.1   Pollen Viability and Germination

Grisales Vásquez et al. (2010) conducted pollen viability tests using a 2% aqueous solution of acetocarmine, and germination tests in 46 uchuva accessions growing in Rionegro, Antioquia State, Colombia. They used a growth medium composed of 1.0, 1.5, or 2.0 g l^{-1} sucrose or glucose each with 0.1% boric acid. Pollen viability among accessions ranged between 54 and 99% after being stored for 2 days at 4 °C. Sucrose plus boric acid (0.1%) was the best solution for pollen germination among the accessions. Germination ranged from 3 to 53% when stored on wetted filter paper at 21 °C for 4 h (Grisales Vásquez et al. 2010). Fernandes da Silva et al. (2017) reported that about 8% pollen germination was obtained at pH 5.4, 27 °C, and 140 g l^{-1} sucrose medium. Addition of calcium nitrate to the medium increased germination to 29%, an addition of boric acid stimulated germination in 23% of the

Fig. 4.1 Scanning electron micrographs showing the structure and exine sculpture of *Physalis peruviana* pollen grains. (**a**) Cluster of pollen grains (scale bar = 10 μm). (**b**) exine sculpture (scale bar = 1 μm). Images from Bhat N, Jeri L, Mipun P, Kumar Y (2018). Systematic studies (micromorphological, leaf architectural, anatomical and palynological) of genus *Physalis* l. (Solanaceae) in northeast India. Plant Arch 18:2229–2238. Images courtesy of Dr. L. Jery. Reproduced with permission

pollen grains. Pollen viability was highest (96%) in anthers collected during the morning soon after anthesis (8:00 am). It decreased progressively through the day such that those collected at 11:00 am were 90% viable, those collected at 2:00 pm were 88%, and those collected in the late afternoon (around 5:00 pm) were 82% viable (Fernandes da Silva et al. 2017). Lagos et al. (2008) however, reported that fresh pollen collected from flowers remained 97% viable after 35–40 days at 13 °C. They found that pollen germinability of fresh pollen grains collected from, five uchuva accessions (UN-49, UN-40, UNPU054, UNPU099, and Kenya) growing in Botana, Nariño State, Colombia ranged between 82 and 98% at 13 °C. The germination rate of pollen collected from these same five accessions growing in El Carmelo, Valle State, Colombia ranged from 67 and 86% at 24 °C (Lagos et al. 2008).

References

Barboza G, Hunziker A, Bernardello G et al (2016) Solanaceae. In: Kadereit J, Bittrich V (eds) Flowering plants, Eudicots, the families and genera of vascular plants 14. Springer, Cham, pp 295–358

Benitez C, Magallanes A (1998) El genero *Physalis* (Solanaceae) de Venezuela. Acta Bot Venezuélica 21:11–42

Bhat N, Jeri L, Mipun P, Kumar Y (2018) Systematic studies (micro-morphological, leaf architectural, anatomical and palynological) of genus *Physalis* L. (Solanaceae) in Northeast India. Plant Arch 18:2229–2238

da Cruz-Barros M, Lima Silva E, Custódio Gasparino E et al (2011) Flora polínica da reserva do parque estadual das Fontes do Ipiranga (São Paulo, Brasil). Hoehnea 38:661–685. https://doi. org/10.1590/S2236-89062007000400008

El-Ghamery A, Khafagi A, Ragab O (2018) Taxonomic implication of pollen morphology and seed protein electrophoresis of some species of Solanaceae in Egypt. Al Azhar Bull Sci 29:43–54

Fernandes da Silva D, Pio R, Vieira Nogueira P et al (2017) Viabilidade polínica e quantificação de grãos de pólen em espécies de fisális. Rev Ciênc Agron 48:365–373

Fernandes da Silva D, Villa F, Sabini da Silva L et al (2019) Caracterização e alterações morfológicas em grãos de pólen de fisális (*Physalis peruviana* L.) (Solanaceae) cultivada sob diferentes espectros luminosos. Colloq Agrariae 15:24–32

Finot V, Marticorena C, Marticorena A (2018) Pollen grain morphology of *Nolana* L. (Solanaceae: Nolanoideae: Nolaneae) and related genera of southern south American Solanaceae. Grana 57:1–41

Ganapathi A, Sudhakaran S, Kulothungan S (1991) The diploid taxon in indian natural populations of *Physalis* L. and its taxonomic significance. Cytologia (Tokyo) 56:283–288

Grisales Vásquez N, Trillos González O, Cotes Torres J, Orozco Orozco L (2010) Estudios de fertilidad de polen en accesiones de uchuva (*Physalis peruviana* L.). Revsita Fac Ciencias Básicas 6:42–51

Halbritter H (2016) *Physalis peruviana*. In: PalDat – a Palynol. database. https://www.paldat.org/pub/Physalis_peruviana/300475. Accessed 1 Jun 2018

Lagos T, Caetano C, Vallejo F et al (2005) Caracterización palinológica y viabilidad polínica de *Physalis peruviana* L. y *Physalis philadelphica* Lam. Agron Colomb 23:55–61. ISSN 0120-9965

Lagos T, Vallejo Cabrera A, Escobar C, Muñoz Flórez J (2008) Biología reproductiva de la uchuva. Acta Agron 57:81–87

Ramírez F (2020) Cocona (*Solanum sessiliflorum* Dunal) reproductive physiology: a review. Genet Resour Crop Evol 67:293–311

Ramírez F (2021) Notes about Lulo (*Solanum quitoense* Lam.): an important South American underutilized plant. Genet Resour Crop Evol 68:93–100. https://doi.org/10.1007/s10722-020-01059-3

Ramírez F, Davenport TL (2020) The development of lulo plants (*Solanum quitoense* Lam. var. *septentrionale*) characterized by BBCH and landmark phenological scales. Int J Fruit Sci 20:562–585. https://doi.org/10.1080/15538362.2019.1613470

Ramírez F, Kallarackal J (2019) Tree tomato (*Solanum betaceum* Cav.) reproductive physiology: a review. Sci Hortic (Amsterdam) 248:206–215

Ramírez F, Fischer G, Davenport TL et al (2013) Cape gooseberry (*Physalis peruviana* L.) phenology according to the BBCH phenological scale. Sci Hortic (Amsterdam) 162:39–42. https://doi.org/10.1016/j.scienta.2013.07.033

Ramírez F, Kallarackal J, Davenport TL (2018) Lulo (*Solanum quitoense* Lam.) reproductive physiology: a review. Sci Hortic (Amsterdam) 238:163–176. https://doi.org/10.1016/j.scienta.2018.04.046

Chapter 5
Pollination

Abstract Uchuva plants rely on both self- and cross-pollination. The bell-shaped flowers are pollinated by wind and insects. *Apis mellifera* has been assumed to be an effective pollinator of field-grown uchuva plants. Bumblebees visiting uchuva plants comprise *Bombus impatiens* and *Bombus atratus*. Other floral visitors include *Xylocopa* sp., *Bombus* sp., the common house fly, Diptera species, stingless bees, wasps, beetles, hemipterans, butterflies, earwigs, and ants. Insect visitor diversity has been observed to be higher in places devoid of insecticide or fungicide applications.

Pollination is required for fruit set in most plants. Nearly every known pollination mechanism occurs in the family, Solanaceae. Pollen deposition can be mediated by bees, birds, moths, butterflies, flies, and bats (Lagos et al. 2008; Barboza et al. 2016). Tropical, solanaceous plants, such as lulo (Ramírez et al. 2018; Ramírez and Davenport 2020; Ramírez 2021), tree tomato (Ramírez and Kallarackal 2019), cocona (Ramírez 2020), and uchuva (Ramírez et al. 2013; Fernandes da Silva et al. 2017), rely on both self- and cross-pollination. Both insects and wind appear to play important roles in pollen transfer and deposition (Gupta and Roy 1981; McCain 1993; Chautá-Mellizo et al. 2012; Singh et al. 2014). Gupta and Roy (1981) found that about 85% of the fruit set in India occurred through self-pollination. Insects and wind easily pollinate the yellow, bell-shaped flowers. Insect pollinators, such as bees, generally appear to help in fruit set (National Research Council 1989), but more research is warranted to fully understand their significance. Whereas pollination is not a problem, inconsistency in fruit size is (McCain 1993).

5.1 Floral Visitors and Insect Pollination

The European honeybee, *Apis mellifera*, was assumed to be an effective pollinator of field-grown uchuva plants in Tena, Colombia and in in potted plants in Ithaca, New York (Chautá-Mellizo et al. 2012). Bumblebees, *Bombus impatiens*, were also

© The Author(s), under exclusive license to Springer Nature   71
Switzerland AG 2021
F. Ramírez, T. L. Davenport, *Uchuva (Physalis peruviana L.) Reproductive Biology*, https://doi.org/10.1007/978-3-030-66552-4_5

observed to visit and possibly pollinate the potted greenhouse uchuva plants grow-
ing in New York. The mass and size of fruits were increased by the occurrence of
Apis mellifera and *Bombus impatiens* (Table 5.1)(Chautá-Mellizo et al. 2012).
Another bumblebee, *Bombus atratus*, is thought to be a pollinator of uchuva in
Colombia (Camelo et al. 2004). Mosquera (2002) reported that the honeybee was
the main floral visitor and possible pollinator of uchuva in Botana and Obonuco,
Nariño State, Colombia. The bees were observed to visit the same flower as many
as eight times within an hour. They were active at 15 °C collecting nectar in the
morning and pollen when available during the afternoon. Other floral visitors
included *Xylocopa* spp., *Bombus* spp., the common house fly, and Diptera species
(Table 5.1) (Mosquera 2002). *Apis dorsata* is a floral visitor of uchuva in Pakistan
(Ali et al. 2017). *Apis mellifera*, *Bombus terrestris*, *Bombus* spp., and stingless bees
have been observed visiting uchuva flowers in Bogotá, Colombia (Ramírez Pers.
Obs.). Honeybees make short 3–5-s visits to flowers while collecting pollen in
Bogotá, and Duitama, Colombia (Fig. 5.1). Stingless bees were observed spending
long periods of at least 10 min within flowers before the onset of a rainy period in
Bogotá, Colombia (Figs. 5.2, 5.3, 5.4, 5.5 and 5.6). They were also observed to col-
lect and carry pollen on their hind legs (Fig. 5.7). Other potential pollinators include
Bombus spp. and *Xylocopa* spp. in Colombia (Chautá-Mellizo et al. 2012). Wasps
and dipterans have been observed visiting uchuva flowers in Chipaque, Cundinamarca
State, Colombia (Figs. 5.8 and 5.9).

Uchuva flowers are visited by arthropods other than bees, bumblebees, stingless
bees and wasps. Beetles, hemipterans, and ants have also been observed visiting
flowers in Bogotá, Colombia (Figs. 5.10, 5.11, 5.12 and 5.13). Hemipterans have
been observed walking on the petals and stamens. Ants were observed consuming
nectar inside the corolla base (Ramírez, personal observation) (Figs. 5.14 and 5.15).
Cream-colored spiders resembling the color of the petals were observed camou-
flaged in the corolla at two locations in Colombia (Figs. 5.16 and 5.17). They appar-
ently stalked prey by remaining immobile within the flower for hours awaiting a
floral visitor (Ramírez personal observation). Earwigs (Dermaptera; Forficulidae)
were also observed visiting flowers in Nariño, Colombia, but their role as potential
pollinators is unclear (Mosquera 2002). Butterflies have been observed visiting
uchuva flowers in Egypt (Table 5.1) (Afsah 2015). Their pupae have been observed
on uchuva leaves in Colombia (Fig. 5.18). Many investigators assume incorrectly
that pollination is taking place just because insects are visiting the flowers, but they
have not determined if pollen deposition or egg fertilization actually occurred as
was demonstrated by Ramírez and Davenport (2013) in apple and Davenport (2019)
in avocado.

Insect visitor number and diversity have been observed to be higher in places
devoid of insecticide or fungicide applications compared to locations were pesti-
cides are frequently applied, both of which occur around Bogotá and Chipaque,
Cundinamarca State, Colombia. A commercial uchuva grower farming near Tuta,
Boyacá State, has noted fewer visiting insects in his orchard when using pesticides
(F. Ramírez, Personal Communication).

Table 5.1 Floral visitors of uchuva

Common name	Scientific name	Country	References
Bees			
Honeybee	*Apis mellifera*	Colombia	Chautá-Mellizo et al. (2012)
Honeybee	*Apis mellifera*	Bogotá, Colombia	Ramírez pers. obs.
Honeybee	*Apis mellifera*	Botana and Obonuco, Nariño State, Colombia	Mosquera (2002)
Bee	*Apis dorsata*	Pakistan	Ali et al. (2017)
Carpenter bees	*Xylocopa* spp.	Botana and Obonuco, Nariño State, Colombia	Mosquera (2002)
Carpenter bees	*Xylocopa* spp.	Colombia	Chautá-Mellizo et al. (2012)
Stingless bees		Bogotá, Colombia	Ramírez pers. obs.
Bumblebees			
Bumblebees	*Bombus impatiens*	Cornell/USA	Chautá-Mellizo et al. (2012)
Bumblebees	*Bombus atratus*	Colombia	Camelo et al. (2004)
Bumblebees	*Bombus atratus*	Bogotá, Colombia	Ramírez pers. obs.
Bumblebees	*Bombus terrestris*	Bogotá, Colombia	Ramírez pers. obs.
Bumblebees	*Bombus* spp.	Bogotá, Colombia	Ramírez pers. obs.
Bumblebees	*Bombus* spp.	Colombia	Chautá-Mellizo et al. (2012)
Bumblebees	*Bombus* spp.	Botana and Obonuco, Nariño State, Colombia	Mosquera (2002)
Flies			
House flies	*Musca domestica*	Botana and Obonuco, Nariño State, Colombia	Mosquera (2002)
Flies	Diptera	Botana and Obonuco, Nariño State, Colombia	Mosquera (2002)
Flies	Syrphidae-Diptera	Chipaque, Cundinamarca State, Colombia	Ramírez pers. obs.
Wasp			
Wasps		Bogotá, Colombia	Ramírez pers. obs.
Wasp		Chipaque, Cundinamarca State, Colombia	Ramírez pers. obs.
Ant			
Ants	*Atta* spp.	Bogotá, Colombia	Ramírez pers. obs.
Hemiptera			
Hemipteras		Bogotá, Colombia	Ramírez pers. obs.

(continued)

Table 5.1 (continued)

Common name	Scientific name	Country	References
Spider			
Spiders		Bogotá, Colombia	Ramírez pers. obs.
Spiders		Duitama, Boyacá State, Colombia	Ramírez pers. obs.
Earwigs			
Earwigs	Dermaptera; Forficulidae	Botana and Obonuco, Nariño State, Colombia	Mosquera (2002)
Beetles			
Beetles	Coleoptera	Bogotá, Colombia	Ramírez pers. obs.
Butterflies			
Butterflies	Lycaenidae	Egypt	Afsah (2015)
Butterflies	Pieridae	Egypt	Afsah (2015)
Butterflies	Noctuidae	Egypt	Afsah (2015)
Butterfllies	Nymphalidae	Egypt	Afsah (2015)

Fig. 5.1 Honeybee visiting an uchuva flower. Photo taken in Duitama, Boyacá State, Colombia

Fig. 5.2 Stingless bee on horizontal position before rain in uchuva flower. Photo taken in Bogotá, Cundinamarca State, Colombia

Fig. 5.3 Stingless bee hanging from uchuva stamens. Photo taken in Bogotá, Cundinamarca State, Colombia

Fig. 5.4 Stingless bee curls around stamens. Photo taken in Bogotá, Cundinamarca State, Colombia

Fig. 5.5 Stingless be walking on a pubescent leaf of uchuva. Photo taken in Bogotá, Cundinamarca State, Colombia

Fig. 5.6 Stingless bee on anthers in uchuva flower. Photo taken in Bogotá, Cundinamarca State, Colombia

Fig. 5.7 Stingless bee with uchuva pollen grains on hind legs. Photo taken in Bogotá, Cundinamarca State, Colombia

Fig. 5.8 Wasp visiting uchuva flower. The wasp stayed in the same position for about 1 min. Photo taken in Chipaque, Cundinamarca State, Colombia

Fig. 5.9 Syrphidae, Diptera visiting an uchuva flower. The visit was 3–5 s. Photo taken in Chipaque, Cundinamarca State, Colombia

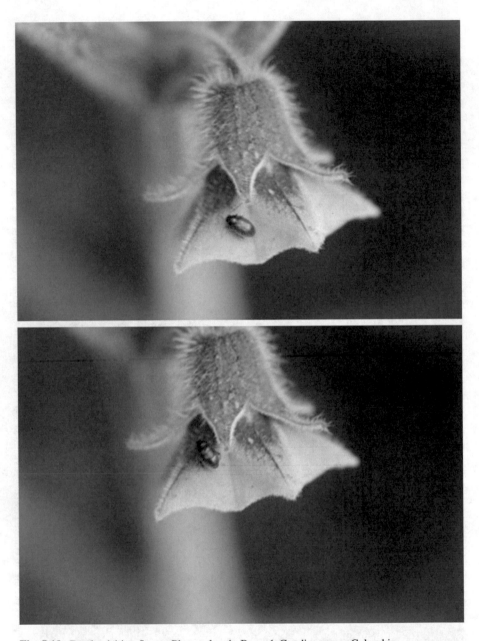

Fig. 5.10 Beetle visiting flower. Photo taken in Bogotá, Cundinamarca, Colombia

Fig. 5.11 Colorful beetle on the underside of a leaf near an uchuva flower. Photo taken in Chipaque, Cundinamarca Sate, Colombia

Fig. 5.12 Hemipteran visiting uchuva flower. Photo taken in Bogotá, Cundinamarca State, Colombia

Fig. 5.13 Close-up view of hemipteran visiting uchuva flower. Note pollen grains evenly spread on maculations. Photo taken in Bogotá, Cundinamarca State, Colombia

Fig. 5.14 Ant feeding on nectar in an uchuva flower. Photo taken in Bogotá, Cundinamarca State, Colombia

Fig. 5.15 Ant crawling inside an uchuva flower, possibly searching for nectar. Photo taken in Bogotá, Cundinamarca State, Colombia

Fig. 5.16 Cream-colored spider guarding uchuva flower. Photo taken in Duitama, Boyacá State, Colombia

Fig. 5.17 Cream-colored nestled inside uchuva flower. Photo taken in Bogotá, Cundinamarca State, Colombia

Fig. 5.18 Butterfly pupa on uchuva leaf. Photo taken in Bogotá, Cundinamarca State, Colombia

References

Afsah AFE (2015) Survey of insects & mite associated cape gooseberry plants (*Physalis peruviana* L.) and impact of some selected safe materials against the main pests. Ann Agric Sci 60:183–191. https://doi.org/10.1016/j.aoas.2015.04.005

Ali M, Sajjad A, Saeed S (2017) Yearlong association of *Apis dorsata* and *Apis florea* with flowering plants: planted forest vs. agricultural landscape. Sociobiology 64:18–25. https://doi.org/10.13102/sociobiology.v64i1.995

Barboza G, Hunziker A, Bernardello G et al (2016) Solanaceae. In: Kadereit J, Bittrich V (eds) Flowering plants, Eudicots, the families and genera of vascular plants 14. Springer, Cham, pp 295–358

Camelo L, Díaz L, Cure J, Almanza M (2004) Morfología floral de la uchuva y comportamiento de visitas de la especie de abejorros *Bombus atratus*. In: XXXI Congreso de la Sociedad Colombiana de Entomología SOCOLEN. Bogotá

Chautá-Mellizo A, Campbell SA, Bonilla MA et al (2012) Effects of natural and artificial pollination on fruit and offspring quality. Basic Appl Ecol 13:524. https://doi.org/10.1016/j.baae.2012.08.013

Davenport TL (2019) Cross- vs. self-pollination in 'Hass' avocados growing in coastal and inland orchards of Southern California. Sci Hortic (Amsterdam) 246:307–316. https://doi.org/10.1016/J.SCIENTA.2018.10.051

Fernandes da Silva D, Pio R, Vieira Nogueira P et al (2017) Viabilidade polínica e quantificação de grãos de pólen em espécies de fisális. Rev Ciênc Agron 48:365–373

Gupta S, Roy S (1981) The floral biology of cape gooseberry (*Physalis peruviana* Linn; Solanaceae, India). Indian J Agric Sci 51:353–355

Lagos T, Vallejo Cabrera A, Escobar C, Muñoz Flórez J (2008) Biología reproductiva de la uchuva. Acta Agron 57:81–87

McCain R (1993) Goldenberry, passionfruit and white sapote: potential fruits for cool subtropical areas. In: Janick J, Simon J (eds) New crops. Wiley, New York, pp 497–486

Mosquera C (2002) Polinizacion entomofila de la uvilla (*Physalis peruviana* L.). Rev Cienc Agrícolas 19:140–158

National Research Council (1989) Lost crops of the Incas: little-known plants of the Andes with promise for worldwide cultivation. National Academy Press, Washington, DC

Ramírez F (2020) Cocona (*Solanum sessiliflorum* Dunal) reproductive physiology: a review. Genet Resour Crop Evol 67:293–311

Ramírez F (2021) Notes about Lulo (*Solanum quitoense* Lam.): an important South American underutilized plant. Genet Resour Crop Evol 68:93–100. https://doi.org/10.1007/s10722-020-01059-3

Ramírez F, Davenport TL (2013) Apple pollination: a review. Sci Hortic 162:188–203

Ramírez F, Davenport TL (2020) The development of lulo plants (*Solanum quitoense* Lam. var. *septentrionale*) characterized by BBCH and landmark phenological scales. Int J Fruit Sci 20:562–585. https://doi.org/10.1080/15538362.2019.1613470

Ramírez F, Kallarackal J (2019) Tree tomato (*Solanum betaceum* Cav.) reproductive physiology: a review. Sci Hortic (Amsterdam) 248:206–215

Ramírez F, Fischer G, Davenport TL et al (2013) Cape gooseberry (*Physalis peruviana* L.) phenology according to the BBCH phenological scale. Sci Hortic (Amsterdam) 162:39–42. https://doi.org/10.1016/j.scienta.2013.07.033

Ramírez F, Kallarackal J, Davenport TL (2018) Lulo (*Solanum quitoense* Lam.) reproductive physiology: a review. Sci Hortic (Amsterdam) 238:163–176. https://doi.org/10.1016/j.scienta.2018.04.046

Singh D, Lal S, Ahamed N, Pal A (2014) Genetic diversity, heritability, genetic advance and correlation coefficient in cape gooseberry (*Physalis peruviana*) under temperate environment. Curr Hortic 2:15–20

Chapter 6
Fruit Morphology

Abstract The ovoid uchuva fruit is an aromatic, juicy, golden-colored, sweet berry. Size differences have been associated with different growing regions and/or cultivar selections. It contains 100–300 small seeds in the orange, fleshy pulp. The seeds have been described as round or reniform, flat, or discoid. Uchuva fruits are covered by the calyx, which is later described as the husk once it desiccates upon fruit maturity. The husk protects the fruit from extreme climatic events, diseases, mechanical damage, and from predation by insects, birds, and mammals.

Morphologically, the uchuva fruit common in South America is a juicy, yellow-orange, aromatic, sweet berry, ovoid or balloon in shape, 0.8–3.5 cm in diameter and 4–10 g in weight (Figs. 6.1, 6.2 and 6.3) (Fischer 2000; Leiva-Brondo et al. 2001; Galvis et al. 2005; Ligarreto et al. 2005; Ramírez et al. 2013; Balaguera-López et al. 2014). This range in size could be associated with different growing regions and / or accession selection. Yıldız et al. (2015) reported that uchuva fruit length ranged from 14 to 20 mm in Turkey. Its diameter was 14–21 mm. Fruit mass ranged from 2.7 to 3.1 g. Fruit density, porosity, and fruit hardness were 462.3 kg/m^3, 53.6%, and 8 N, respectively. Stomata occur primarily on the distal surface of fruit and are less abundant near the stem end (Perea et al. 2010).

The uchuva berry contains 100–300 small seeds randomly immersed within the orange, fleshy pulp (Figs. 6.4 and 6.5) (Duarte and Paull 2015; Bhat et al. 2018). The seeds have been described as round or reniform (Ligarreto et al. 2005), flat (Benitez and Magallanes 1998), or discoid (Bhat et al. 2018) and about 1.7–2.2 mm in length and 1.5–2 mm in diameter (Fig. 6.6). The seed surface is reticulate-foveate and lustrous yellow to pale brown (Benitez and Magallanes 1998; Bojňanský and Fargašová 2007; Bhat et al. 2018). Axelius (1992) found that the testa pattern on the seeds has anticlinal walls that are deeply undulated with small edges, and are deeply sunken between the wall thickenings.

Uchuva fruits are covered by the calyx or husk, once it desiccates upon fruit maturity (Fig. 6.7). The husk protects the fruit from extreme climatic events, diseases, mechanical damage, and from predation by insects, birds, and mammals

F. Ramírez, T. L. Davenport, *Uchuva (Physalis peruviana L.) Reproductive Biology*, https://doi.org/10.1007/978-3-030-66552-4_6

Fig. 6.1   Ripe uchuva fruit. Berry viewed form top. Photo taken in Bogotá, Cundinamarca State, Colombia

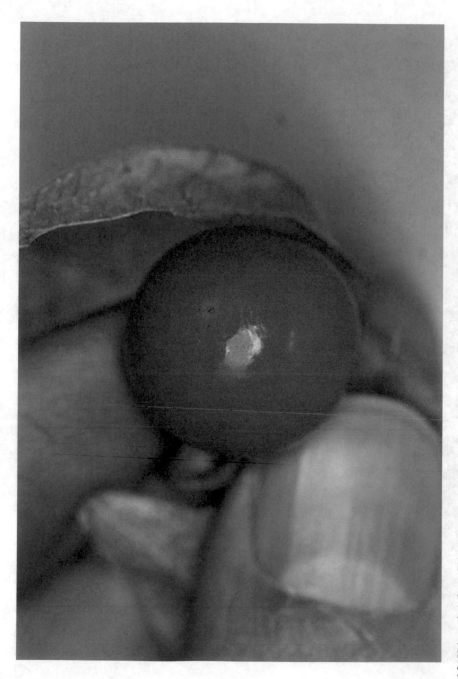

Fig. 6.2 Uchuva fruit from Colombia accession. Photo taken in Bogotá, Cundinamarca State, Colombia

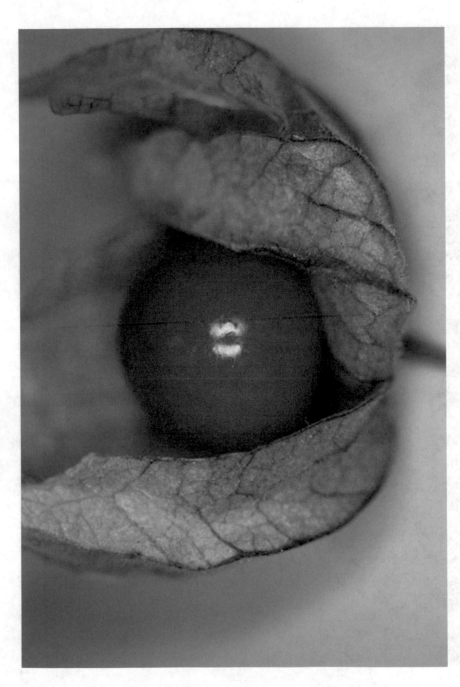

Fig. 6.3 Detail of Colombia accession with husk. Photo taken in Bogotá, Cundinamarca State, Colombia

Fig. 6.4 Detail of uchuva pulp and seeds. Photo taken in Bogotá, Cundinamarca State, Colombia

Fig. 6.5 Pulp and seeds in Colombia uchuva accession. Photo taken in Bogotá, Cundinamarca State, Colombia

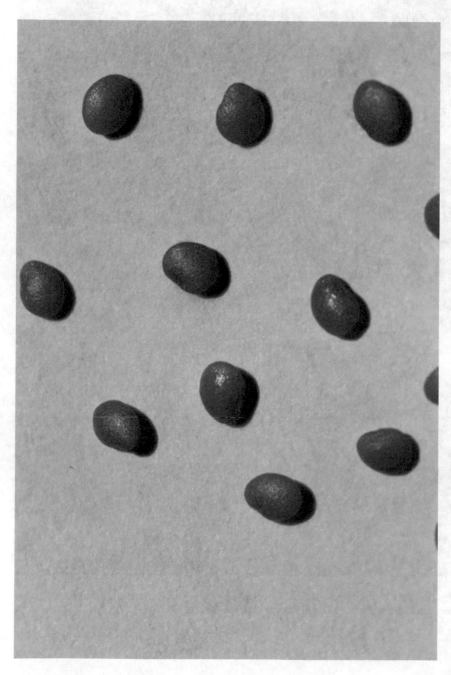

Fig. 6.6 Detail of uchuva seeds. Photo taken in Bogotá, Cundinamarca State, Colombia

Fig. 6.7 Uchuva husk in the Colombia accession. Upper photo: Immature husk. Lower photo: Near mature husk. Photos taken in Bogotá, Cundinamarca, Colombia

(Fischer et al. 2011). Within the husk, the glandular tissue located at the inner base of the calyx produces a resin that coats the fruit (Fischer et al. 2011). The mature fruit's skin is yellow to yellow-orange, shiny and smooth (Galvis et al. 2005).

References

Axelius B (1992) Testa patterns in some species of *Physalis* L and some other genera in the tribe Solaneae (Solanaceae). Int J Plant Sci 153:488–502. https://doi.org/10.1086/297055

Balaguera-López H, Ramírez Sanabria L, Herrera Arévalo A (2014) Fisiología y bioquímica del fruto de uchuva (*Physalis peruviana* L.) durante la maduración y poscosecha. In: Carvalho C (ed) Uchuva (*Physalis peruviana* L.) fruta Andina para el mundo. Limencop S.L, Alicante, pp 113–131

Benitez C, Magallanes A (1998) El genero *Physalis* (Solanaceae) de Venezuela. Acta Bot Venezuélica 21:11–42

Bhat N, Jeri L, Mipun P, Kumar Y (2018) Systematic studies (micro-morphological, leaf architectural, anatomical and palynological) of genus *Physalis* L. (Solanaceae) in Northeast India. Plant Arch 18:2229–2238

Bojňanský V, Fargašová A (2007) Atlas of seeds and fruits of central and east-European flora. Springer, Dordrecht

Duarte O, Paull R (2015) Exotic fruits and nuts of the new world. CABI, Wallingford

Fischer G (2000) Crecimiento y desarrollo. In: Flórez V, Fischer G, Sora A (eds) Producción, poscosecha y exportación de la uchuva (*Physalis peruviana* L.). Universidad Nacional de Colombia, Unibiblos, Bogotá, pp 9–26

Fischer G, Herrera A, Almanza PJ (2011) Cape gooseberry (*Physalis peruviana* L.). In: Yahia EM (ed) Postharvest biology and technology of tropical and subtropical fruits. Woodhead Publishing, Oxford, pp 374–397

Galvis J, Fischer G, Gordillo O (2005) Cosecha y poscosecha de la uchuva. In: Fischer G, Miranda D, Piedrahíta W, Romero J (eds) Avances en cultivo, poscosecha y exportación de la uchuva (*Physalis peruviana* L.) en Colombia. Unibiblos, Bogotá, pp 165–190

Leiva-Brondo M, Prohens J, Nuez F (2001) Genetic analyses indicate superiority of performance of cape gooseberry (*Physalis peruviana* L.) hybrids. J New Seeds 3:71–84. https://doi.org/10.1300/J153v03n03_04

Ligarreto G, Lobo M, Correa M (2005) Recursos genéticos del género *Physalis* en Colombia. In: Fisher G, Miranda D, Piedrahita W, Romero J (eds) Avances en cultivo, poscosecha y exportación de la uchuva (*Physalis peruviana* L.) en Colombia. Universidad Nacional de Colombia, Unibiblos, Bogotá, pp 9–26

Perea M, Rodríguez N, Fischer G et al (2010) Uchuva: *Physalis peruviana* L. (Solanaceae). In: Perea M, Matallana L, Tirado A (eds) Biotecnología aplicada al mejoramiento de los cultivos de frutas tropicales. Universidad Nacional de Colombia, Bogotá, pp 466–490

Ramírez F, Fischer G, Davenport TL et al (2013) Cape gooseberry (*Physalis peruviana* L.) phenology according to the BBCH phenological scale. Sci Hortic (Amsterdam) 162:39–42. https://doi.org/10.1016/j.scienta.2013.07.033

Yıldız G, İzli N, Ünal H, Uylaşer V (2015) Physical and chemical characteristics of goldenberry fruit (*Physalis peruviana* L.). J Food Sci Technol 52:2320–2327. https://doi.org/10.1007/s13197-014-1280-3

Chapter 7
Propagation

Abstract Planting seeds taken from fresh fruit is the most widely used propagation method for uchuva plants. Uchuva seed germination rates range from 85 to 90%. Seedling emergence occurs in 10–15 days after planting. Propagation by cuttings has been used for fast and uniform plant production with the same genetic makeup. Plant growth regulators have been used to assist in propagation of cuttings. *In vitro* culture has also been an effective mode of plant propagation, particularly in combination with plant growth promoting substances.

7.1 Seed Germination

Planting seeds from fruit is the most widely used method for uchuva plants (Miranda 2005; Fischer and Miranda 2012). Uchuva seeds have a high germination rate ranging from 85 to 90%. Miranda (2005) and Duarte and Paull (2015) reported that seedling emergence has been associated with seed quality, substrate type, and usually takes from 10 to 15 days to occur. Seeds are usually germinated in plastic germination trays. During the seedling stage, uchuva plants are transferred to one-pound, plastic bags filled with loamy soil. They remain in the bags for 30–60 or more days after sowing, depending on seed vigor, substrate type and composition (Flórez 1986; Angulo 2000; Miranda 2005).

   Upon germination, the radicle emerges from the seed (Figs. 7.1, 7.2 and 7.3), and the hypocotyl subsequently breaks through the seed coat forming a hook to protect the apical meristem as it forces its way through the soil (Figs. 7.4 and 7.5). The hooked hypocotyl (the elongating stem beneath the two cotyledons) pushes up through the soil into the light (Figs. 7.6 and 7.7) (Ramírez et al. 2013). The hypocotyl hook then straightens to display the two cotolydonary leaves (Figs. 7.8 and 7.9). Once the hypocotyl straightens, the two cotyledons start to open until reaching a horizontal position (Figs. 7.10, 7.11 and 7.12). Nuñes et al. (2018) found that uchuva seeds incubated on wetted Germitest paper at controlled constant temperatures of 27 °C and 32 °C exhibited high germination rates of 98% and 96%, respectively. Germination, however, was reduced when seeds were incubated at 12, 17, 22, and

© The Author(s), under exclusive license to Springer Nature
Switzerland AG 2021
F. Ramírez, T. L. Davenport, *Uchuva (Physalis peruviana L.) Reproductive
Biology*, https://doi.org/10.1007/978-3-030-66552-4_7

Fig. 7.1 Radicle breaks through seed coat. Photo taken in Bogotá, Cundinamarca State, Colombia

Fig. 7.2 Emerged radicle with developing root hairs. Photo taken in Bogotá, Cundinamarca State, Colombia

Fig. 7.3 Radicle elongates ageotropically to enter the soil. Note the fine root hairs. Photo taken in Bogotá, Cundinamarca State, Colombia

Fig. 7.4 Radicle visible as hypocotyl emerges from the seed coat. Photo taken in Bogotá, Cundinamarca State, Colombia

Fig. 7.5 Inverted U-turn hypocotyl breaking though soil. Photo taken in Bogotá, Cundinamarca State, Colombia

Fig. 7.6 The hooked hypocotyl emerges from soil exposing cotyledons to sunlight. Note the seed coat loosely attached to tips of the cotyledons. Photo taken in Bogotá, Cundinamarca State, Colombia

Fig. 7.7 Seed coat detaches from cotyledons. Photo taken in Bogotá, Cundinamarca State, Colombia

Fig. 7.8 Cotyledond turn upwards. Photo taken in Bogotá, Cundinamarca State, Colombia

Fig. 7.9 Cotyledons in an upward position. Note how they start to open. Photo taken in Bogotá, Cundinamarca State, Colombia

Fig. 7.10 Cotyledons nearly horizontal. Note the pubescences. Photo taken in Bogotá, Cundinamarca State, Colombia

Fig. 7.11 Top view of cotyledons. Photo taken in Bogotá, Cundinamarca State, Colombia

Fig. 7.12 Cotyledons in horizontal orientation. Note the pubescences. Photo taken in Bogotá, Cundinamarca State, Colombia

37 °C, constant. They determined that photoperiods longer than 8 h of light per day in the incubators were ideal for seed germination, but they were not clear whether this benefit is a result of a photoperiodically controlled event or simply longer availability of radiant warmth during the programed daylight hours. Almanza (2000) found that 90% of the Kenyan ecotype germinated by 10 days after planting at 20 °C and 80% relative humidity. Orozco Orozco et al. (2010) obtained 70–80% seed germination in ten Colombian uchuva ecotypes 15 days after sowing at 16 °C.

Criollo and Ibarra (1992) reported best germination results (73.4%) occurred in Colombian uchuva seeds derived from ripe fruits, whereas seeds from green immature fruits had the lowest germination rate (41.6%). Seeds from both fruit stages were sown in coarse river sand and maintained at average greenhouse temperature conditions of 22 °C. Criollo and Upegui (2005) observed 89% germination after 15 days in seeds harvested from 90-day-old, mature Colombian uchuva fruits. Fresh uchuva seeds were germinated on filter paper in incubator conditions at 25 °C and 95% relative humidity. In contrast, Diniz et al. (2019) found that 75% of uchuva seeds germinated 105 days after planting. Seeds were obtained from 90-day-old, mature fruits in Brazil. Although seeds were similarly germinated on moist blotting paper and under constant incubator conditions of 25 °C and 8-h photoperiod as those described by Criollo and Upegui (2005), the time to germinate was greatly different suggesting an unknown variable that may have been involved. Mazorra et al. (2003) reported 95% seed germination at 15 °C and 80% relative humidity obtained from 45-day-old fresh uchuva fruits. Seed germination was 35% in 30-day-old fruits. Seeds were germinated on moist filter paper.

Uchuva seed germination rate is reduced after storage. Betancourt et al. (2008) reported that seed stored for 7 days at 11% relative humidity provided 60% germination, whereas seeds stored below 4% relative humidity did not germinate. Best results, 98% seed germination, were obtained at approximately 10% relative humidity. de Souza et al. (2016) reported storage of germinable uchuva seeds up to a year using complicated protocols involving chemical dehydration and liquid nitrogen (−196 °C).

Although uchuva seed germination was lowered by high salinity (de Souza et al. 2016), Miranda et al. (2010) reported that seeds exposed to 30 mM NaCl salt on a filter paper substrate had a germination percentage of 96.4, which was the same as the control treatment without salt (97.6% germination). Seeds exposed to 120 mM NaCl had a lower germination rate of 62.5%. Uchuva has, thus, been classified by these authors as moderately tolerant to sodium. Similarly, germination rate decreased with increasing NaCl concentration, but salinity tended to affect seed emergence and seedling growth more than seed germination (Yildirim et al. 2011). No seed emergence occurred at 90, 120, and 180 mM NaCl.

Díaz et al. (2012) examined uchuva seed germination in six plant growth substrates, Brazilian coco peat, fine river sand, German blond peat moss, Canadian peat moss with perlite, a 50:50 mix of German blond peat moss with rice hulls, and a 50:50 mix of coco peat and rice hulls under greenhouse conditions of 22 °C day/18 °C night. Seed germination started after 13 days. Canadian peat moss with perlite was the best substrate for germination with 123 germinated seeds per

128-cell seedling tray 38 days after sowing, whereas fine river sand had the lowest seed germination (70 seedlings per 128-cell tray) 38 days after sowing.

Shade nets colored white, blue, red, and black, all with 50% shading, have been used to evaluate germination response of uchuva seeds to light quality (Fernandes Da Silva et al. 2016). Emergence speed index was best for germinating uchuva seeds and three other *Physalis* species shaded by white and red nets. Pellizzaro et al. (2019) germinated uchuva seeds in incubators at a constant temperature of 25 °C and at diurnally fluctuating temperatures from 25 to 30 °C. Seeds were also exposed to blue, red, and white light conditions and total darkness. Seeds were germinated on moist paper. Best germination results were obtained at 25 °C coupled with any color of light. Seed germination was significantly reduced in total darkness at 25 °C.

7.2   Vegetative Propagation

7.2.1   Cuttings

Uchuva propagation by cuttings is recommended for fast and uniform plant production while preserving the same genetic features as those of the mother plant (Perea et al. 2010). Cutting-propagated plants produce fruits earlier and larger than those obtained from seed (Klinac 1986; Fischer and Miranda 2012). Uchuva cuttings ranging from 10 to 25 cm long and treated with a root-stimulating plant-growth regulator have provided more success than air layering in India (Ali and Singh 2013). In Hawaii, uchuva stem cuttings are made from shoots cut 6–8 inches in length, treated with a growth regulator to induce rooting, and planted in flats kept in a moist, shaded area. They are ready for transplant after 2–3 weeks (Chia et al. 1987).

Production of vigorous, long, and uniform roots is essential for successful propagation by cuttings (Almanza 2000). This can be achieved by using auxin-like, root-growth-promoting substances. For example, Sandhu et al. (1989) reported that greatest root growth (45%) occurred at a concentration of 250 ppm indole butyric acid (IBA). Verhoeven (1991) used 1% IBA to stimulate rooting in stem cuttings and observed earlier flowering. Zeist et al. (2015) evaluated propagation of uchuva cuttings using a nutrient medium, Mecplant® in combination with other substrates in a hydroponic system. They preferred using propagation by cuttings due to its ease to produce a strong root system.

López Acosta et al. (2008) evaluated root length, fresh, dry weight of roots, shoot length, and fresh weight of uchuva cuttings at Universidad Nacional de Colombia, Bogotá, Cundinamarca State. Cuttings used in the experiments were excised from the upper, middle, and lower section of uchuva stems. Each cutting was 16 cm long, bearing two nodes, and a pair of leaves on the upper node. They were placed either in sand, soil plus rice husks (1:1 v/v), or soil, rice husks, and sand (1:1:1 v/v) without a growth regulator at 90% relative humidity and 24 °C. Best results were obtained for cuttings taken from the upper section of stems and rooted in sand. They had the longest root length, highest root dry and fresh matter, and highest percentage of rooting.

Moreno et al. (2009) applied 0, 200, 400, 600, and 800 ppm IBA to 25 cm uchuva cuttings bearing two leaves. The experiment was conducted inside mesh cages at 18 °C and 75% relative humidity at Universidad Pedagógica y Tecnológica de Colombia, Tunja, Boyacá State. Cuttings were planted in 2-liter polyethylene bags. They used two growing substrates, Canadian peat moss, and a 1:1 v/v mixture of black soil and rice husks. Data were collected after 90 days. Best rooting results were obtained in the 800 ppm IBA using peat moss as a substrate. Alvarado-Sanabria and Álvarez-Herrera (2014) applied 0, 800, 1200, and 1600 ppm of IBA to 20-cm-long uchuva cuttings with a pair of leaves in a greenhouse located at the same institution. The mean temperature was 18 °C and 75% relative humidity. The basal end of the cuttings were immersed for 5 min in the IBA solutions and then planted in 2-liter pots filled with Californian peat moss mixed with burned rice husks at a proportion of 3:1 (v/v). They then applied suspensions of the fungus *Trichoderma harzianum*, at rates of 0, $2 \cdot 10^6$, $3 \cdot 10^6$, or $4 \cdot 10^6$ cfu ml^{-1} distributed among the different IBA treated cutting. Best results were obtained in the 800 ppm IBA application plus $3 \cdot 10^6$ cfu ml^{-1}. These cuttings displayed the highest root dry mass, total dry mass and leaf area.

de Oliveira et al. (2015) evaluated the vegetative propagation of uchuva by cuttings and the application of the auxin, IBA, to stimulate rooting. They used 12-cm-long cuttings the basis of which were immersed for 5 min in 0, 400, 800, 1200, 1600 mg l^{-1} IBA. Cuttings were planted in 300 ml plastic cups containing 50% Germina Plant® substrate, 50% sand and were grown in greenhouse conditions in Brazil. The number of shoots, leaves, and shoot length were measured after 20 and 60 days. Furthermore, leaf dry weight and total dry weight, percent survival, root number, and length of the longest root were measured after 60 days. They concluded that it was not necessary to use IBA during uchuva propagation. The results and findings, however, of this investigation were difficult to interpret.

Fischer et al. (1989) reported that propagation by cuttings improved yield over that of seed-germinated plants in Boyacá State, Colombia. Cuttings were rooted in plastic bags and transplanted in the field at a spacing of 0.8 × 1.0 m and later pruned 6–8 main shoots. Despite these success stories, Almanza (2000), de Souza et al. (2016), and Fischer and Miranda (2012) concluded that propagation by cuttings not be recommended for commercial plantings due to its high cost, comparatively poor root development, and susceptibility to wind damage. Klinac (1986) reported that plants propagated from cuttings were less vigorous, more prone to fruit splitting, and had lower soluble solids in fruit than plants propagated from seed.

7.2.2   In vitro *Culture*

In vitro culture of uchuva has been an effective asexual mode of plant propagation when used in combination with plant growth promoting substances including 6-benzylaminopurine (BAP), kinetin, gibberellic acid (GA$_3$), the auxins, naphthaleneacetic acid (NAA), 2,4-dichlorophenoxyacetic acid (2,4-D), indole-3-acetic acid (IAA), and indole-3-butyric acid (IBA). Fischer and Miranda (2012) found that *in vitro*-propagated uchuva plants produced adventitious roots that negatively impacted the plants´ ability to establish anchorage in the soil anchor.

Guney et al. (2016) propagated uchuva using tender, elongating apical shoots as the source of their explants. Explants were grown *in vitro* on basal medium (McCown's woody plant basal salts) that included vitamins, 20 g l^{-1} sucrose, and 7.5 g l^{-1} agar. They were grown in a plant growth chamber maintained at 25 °C under 16-h photoperiod, using cool-white fluorescent lamps (1500 lux) and an 8-h dark cycle. The objective was to regenerate whole, rooted plants using 6-BAP (1, 2, 3 mg l^{-1}) in combination with IBA (0, 0.1, 0.2, 0.4 mg l^{-1}) or NAA (0, 0.1, 0.2, 0.4 mg l^{-1}). Higher auxin concentrations (1 and 2 mg l^{-1} IBA or 1 and 2 mg l^{-1} NAA) without 6-BAP was also applied in an effort to stimulate root formation in the explants. The greatest number of regenerated rooted shoots was obtained in the 2 mg l^{-1} BAP combined with 0.4 mg l^{-1} IBA. The greatest shoot length occurred in the 2 mg l^{-1} BAP without NAA. The highest number of roots occurred in shoots treated with 2 mg l^{-1} and 1 mg l^{-1} NAA. Mascarenhas et al. (2019) established a protocol for uchuva micropropagation through direct organogenesis. They incubated cotyledon, cotyledonary node, hypocotyl, leaf, epicotyl, and root explants in half strength Murashige & Skoog culture medium (MS) (Murashige and Skoog 1962). The media was furthermore supplemented with 0, 2.22, 4.44, 6.66, or 8.88 µM BAP plus 30 g l^{-1} sucrose and 7 g l^{-1} agar. Cultures were maintained in a growth room at 23 °C, 16-h photoperiod, and photosynthetically active radiation of 60 µmol $m^{-2}\,s^{-1}$ supplied by cool-white fluorescent lamps. Best shoot regeneration was obtained in cotyledonary node and leaf explants.

Ramar et al. (2014) reported effective *in vitro* regeneration of uchuva node, internode, and leaf explants. They used MS medium (Murashige and Skoog 1962) with B5 vitamins, BAP (0.5–4.0 mg l^{-1}), GA_3 (0.5–2.5 mg l^{-1}) and 2,4-D (0.5–2.0 mg l^{-1}). Temperature conditions were 25 °C, 55–65% relative humidity, and 16-h photoperiod of 35 mol $m^{-2}\,s^{-1}$ irradiance provided by cool white fluorescent bulbs. Best results, showing the highest numbers of multiple shoots from nodal and internodal explants were obtained in MS medium with 2.0 mg l^{-1} BAP + 1.0 mg l^{-1} GA_3 + 1.0 mg l^{-1} 2,4-D. Leaf explants exhibited the greatest shoot multiplication in MS medium supplemented with 3.0 mg l^{-1} BAP, 1.0 mg l^{-1} GA_3, and 1.0 mg l^{-1} 2,4-D.

Zenkteller (1972) excised 2.5-cm-long leaves from 8-week-old-*in-vitro* cultured seedlings. The leaves were placed on LS medium (Linsmaier and Skoog 1965). Treatments were LS medium fortified: (a) without hormones, (b) with 1 mg l^{-1} kinetin and 1 mg l^{-1} IAA, (c) with 2 mg l^{-1} kinetin and 1 mg l^{-1} IAA, (d) with 4 mg l^{-1} kinetin and 2 mg l^{-1} IAA, (e) with 6 mg l^{-1} kinetin and 2 mg l^{-1} IAA, (f) with 1, 2, 4, and 6 mg l^{-1} kinetin alone, and (g) with 1, 2, 4, and 6 mg l^{-1} IAA alone. The culture media were maintained at 24 °C. Both the 4 mg l^{-1} kinetin plus 2 mg l^{-1} IAA treatment and the 6 mg l^{-1} kinetin plus 2 mg l^{-1} IAA treatment produced a few buds and occasional plantlets after 6 weeks. The 4 mg l^{-1} kinetin alone treatment also produced occasional buds and plantlets.

Torres et al. (1991) excised meristems from apical and lateral buds on 4–6-month-old uchuva plants. They used MS culture medium fortified with 22.2 µM BAP at

25 °C, 16-h photoperiod, and 8-h night. Apical buds developed into plantlets after 2–3 weeks whereas lateral buds produced 5–8 shoots without roots after 4 weeks. Individual shoots were cultured in MS culture medium in addition to 10.42 μM IAA for root and plantlet formation.

Rodríguez et al. (2013) used 2-cm uchuva stem segments for *in vitro* propagation. The agar substrate contained 0, 25, 50, 75, and 100% MS culture medium in addition to 0, 0.5, 1.0, 1.5, and 2.0 mg l^{-1} 6-BAP. They were kept at 25 °C and 16-h photoperiod. Best results were obtained in the 50% MS media fortified with 1,5 mg l^{-1} 6-BAP. This treatment produced the highest number of shoots (3) among all treatments.

Martínez-Montiel et al. (2011) used 3-cm-long uchuva shoot explants for *in vitro* propagation. They used MS agar culture medium supplemented with 0.9 mg l^{-1} thiamine, 0.5 mg l^{-1} folic acid, 0.05 mg l^{-1} biotin, and 30 g l^{-1} sucrose. They were maintained in an incubator set at 26 °C and 16-h day, 8-h night photoperiod. Best rooting results were obtained in 50% MS agar medium. This medium produced the tallest shoots (14.4 cm), longest roots (6 cm), number of buds (3.3), and number of leaves (4.3).

Chaves et al. (2005) examined *in vitro*, uchuva propagation using 1-cm shoots as explants. They used MS agar media and MS and agar media with MS salts reduced by 25% both, supplemented with 0, 0.1, 0.2, or 0.3 mg l^{-1} 6-BAP. The 30-ml, glass tubes containing the explants in agar media were maintained at an illumination flux of 42 μmol m^{-2} s^{-1} and a temperature of 25 °C. Best results with longer shoots were obtained in the 0.3 mg l^{-1} 6-BAP after 21 days.

Jayasree et al. (2005) reported that uchuva shoots developed directly on leaf explants without the normal callus phase when they were placed on MS agar medium containing 6-BAP alone or in combination with the auxins, IAA or NAA. The auxins, 2,4-D, 2,4,5-T, IAA or NAA, alone, stimulated only root growth on the explants. Best results for maximum number of adventitious shoots was obtained in the MS agar medium supplemented with the auxin, IAA plus the cytokinin, 6-BAP. Highest root formation frequency of leaf explants occurred in media supplemented with the auxins 2,4-D or 2,4,5-T.

Santana and Angarita (1997) reported that hypocotyl explants of uchuva were used for callus formation and plant regeneration. They used MS agar medium supplemented with 0–5 ppm 6-BAP and 0 or 1 ppm GA_3 in combination with 0 or 1.0 ppm NAA or 2,4-D at 23 °C and continuous light conditions provided by florescent lamps at a flux of 18–24 mmol m^{-2} s^{-1} or total darkness. The greatest number of regenerated plants was observed on the media supplemented with 1 ppm 6-BAP, 1 ppm GA_3, and 0.5 ppm 2,4-D under continuous light.

Rache Cardenal and Pacheco Maldonado (2013) used uchuva axillary buds as explants for *in vitro* propagation. They used the MS medium maintained at 24 °C and continuous light conditions with an intensity of 70–80 mmol m^{-2} s^{-1}. The MS agar medium was supplemented with the cytokinin, 0.1 mg l^{-1} 6-BAP and the auxin, 0.05 mg l^{-1} IBA. Explants produced 100% rooting. Their protocol was effective to propagate uchuva plants.

Criollo Escobar et al. (2001) used apical buds as explants for *in vitro* propagation of uchuva. They used MS agar medium plus 1.0 ppm GA₃, 0.05 ppm 2,4-D, and 1 ppm 6-BAP maintained at 20 °C and 57% relative humidity. They observed callus formation and growth for 14 weeks. The callus was transferred to six different culture media containing MS medium supplemented with a variety phytohormones and vitamins. All treatments induced callus formation and subsequent vegetative shoots, but only MS coupled with 0.02 ppm NAA produced whole uchuva plants.

Celikli et al. (2017) used 1-cm long uchuva shoot apices as explants for *in vitro* propagation. They used MS medium containing 1 mg l⁻¹ IAA, 30 g l⁻¹ sucrose, and 7 g l⁻¹ agar in addition to 0, 25, 50, 75, and 100 mM NaCl. Cultures were maintained at 25 °C, 16-h photoperiod for 4 weeks. They used fluorescent bulbs with a light intensity of 40 μmol m⁻² s⁻¹. Shoot length was reduced with increasing salt concentration.

Singh et al. (2016) used 1–2-cm uchuva nodal segments as explants for *in vitro* culture. They used MS medium with 0–4.5 mg l⁻¹ BAP and 0–0.09 mg l⁻¹ IBA at 25 °C, and 16-h photoperiod. Best shoot growth results were observed in the 2.5 mg l⁻¹ BAP and 0.05 mg l⁻¹ IBA treatments.

Hernández-Villalobos and Chico-Ruíz (2020) used uchuva hypocotyls and cotyledons as explants for *in vitro* propagation. They used nine combinations of growth regulators between 2,4-D (0.25, 0.5, or 1.0 mg l⁻¹) or 6-BAP (0.1, 0.25, or 1.0 mg l⁻¹) in MS medium with 3% sucrose, 0.5% phytagel, pH 6, and 16-h photoperiod. They concluded that cotyledons were better for root and shoot regeneration than hypocotyls.

References

Ali A, Singh B (2013) Potentials of cape gooseberry (*Physalis peruviana* L): an under-exploited small fruit in India. Asian J Hortic 8:775–777

Almanza J (2000) Propagación. In: Flórez V, Fischer G, Sora A (eds) Producción, poscosecha y exportación de uchuva (*Physalis peruviana* L.). Universidad Nacional de Colombia, Facultad de Agronomía, Bogotá, pp 27–40

Alvarado-Sanabria O, Álvarez-Herrera J (2014) Efecto del ácido indol-3-butírico y *Trichoderma harzianum* Rifai en la propagación asexual de uchuva (*Physalis peruviana* L.). Agron Colomb 32:326–333. https://doi.org/10.15446/agron.colomb.v32n3.45941

Angulo C (2000) Siembra, soporte, poda y fertilización de la uchuva. In: Florez V, Fischer G, Sora A (eds) Cultivo, poscosecha y exportación de la uchuva (*Physalis peruviana* L.). Unibiblos, Universidad Nacional de Colombia, Bogotá, pp 41–50

Betancourt MLB, Piedrahíta KE, Terranova AMP et al (2008) Establecimiento de una colección de trabajo de uchuva del suroccidente colombiano. Acta Agron 57:95–99

Celikli FB, Akkelle P, Onus AN (2017) *In vitro* salinity evaluation studies in golden berry (*Physalis peruviana* L.). Biotechnol J Int 20:1–8

Chaves A d C, Schuch MW, Erig AC (2005) Estabelecimento e multiplicação *in vitro* de *Physalis peruviana* L. Ciênc Agrotecnol 29:1281–1287. https://doi.org/10.1590/s1413-70542005000600024

Chia C, Nishina N, Evans D (1987) Poha. University of Hawaii at Manoa, Honolulu

Criollo Escobar H, Chavez Jurado G, Lagos Burbano T (2001) Evaluación de diferentes medios para la conservación de germoplasma de uvilla (*Physalis peruviana* L.) *in vitro*. Rev Cienc Agrícolas 18:181–190

Criollo H, Ibarra V (1992) Germination of cape gooseberry (*Physalis peruviana* L.) under different degrees of maturity and storage times. Acta Hortic 310:183–188

Criollo E, Upegui P (2005) Determinación de la madurez fisiológica de semillas de uvilla (*Physalis peruviana* L.). Rev Cienc Agrícolas 22:1–13

de Oliveira JAR, Koefender J, Manfio CE et al (2015) Tipos de estacas e uso de AIB na propagação vegetativa de fisális. Rev Agro@Mbiente On-Line 9:342. https://doi.org/10.18227/1982-8470ragro.v9i3.2571

de Souza CLM, de Souza MO, Oliveira RS et al (2016) *Physalis peruviana* seed storage. Rev Bras Eng Agrícola Ambient 20:263–268. https://doi.org/10.1590/1807-1929/agriambi.v20n3p263-268

Díaz LA, Fischer G, Pulido SP (2012) La fibra de coco como sustituto de la turba en la obtención de plántulas de uchuva (*Physalis peruviana* L.). Rev Colomb Cienc Hortícolas 4:143–152. https://doi.org/10.17584/rcch.2010v4i2.1236

Diniz FO, Novembre AD d LC, Diniz FO, Novembre AD d LC (2019) Maturation of *Physalis peruviana* L. seeds according to flowering and age of the fruit. Rev Ciênc Agron 50:447–457. https://doi.org/10.5935/1806-6690.20190053

Duarte O, Paull R (2015) Exotic fruits and nuts of the new world. CABI, Wallingford

Fernandes Da Silva D, Pio R, Dória J et al (2016) The production of *Physalis* spp. seedlings grown under different-colored shade nets. Acta Sci 38:257–263. https://doi.org/10.4025/actasciagron.v38i2.27893

Fischer G, Miranda D (2012) Uchuva (*Physalis peruviana* L.). In: Fischer G (ed) Manual para el cultivo de frutales en el trópico. Produmedios, Bogotá, pp 851–873

Fischer G, Buitrago M, Lüdders P (1989) *Physalis peruviana* L. – cultivation and investigation in Colombia. Erwerbs-obstbau 32:229–232

Flórez L (1986) Tecnología del cultivo de la uchuva (*Physalis peruviana* L.). In: Memorias Primer Curso Nacional de Uchuva. UPTC Tunja, Tunja, pp 6–15

Guney M, Kafkas S, Kefayati S et al (2016) *In vitro* propagation of *Physalis peruviana* (L.) using apical shoot explants. Acta Sci Pol Hortorum Cultus 15:109–118

Hernández-Villalobos K, Chico-Ruíz J (2020) Inducción de brotes y raíces en hipocotilos y cotiledones de *Physalis peruviana* l. utilizando 6-bencilaminopurina y 2,4-diclorofenoxiacético. Rev Investig Altoandinas 22:86–94

Jayasree A, Seetaram A, Venugopal Rao K et al (2005) *In vitro* response of leaf explants of cape goose berry (*Physalis peruviana* L.). Plant Cell Biotechnol Mol Biol 6:115–120

Klinac DJ (1986) Cape gooseberry (*Physalis peruviana*) production systems. N Z J Exp Agric 14:425–430. https://doi.org/10.1080/03015521.1986.10423060

Linsmaier EM, Skoog F (1965) Organic growth factor requirements of tobacco tissue cultures. Physiol Plant 18:100–127

López Acosta J, Tenjo Guío N, Miranda Lasprilla D (2008) Propagación de uchuva (*Physalis peruviana* L.) mediante diferentes tipos de esquejes y sustratos. Rev Fac Nac Agron 61:4347–4357

Martínez-Montiel OM, Pastelín-Solano MC, Ventura-Zapata E et al (2011) Alargamiento y enraizamiento de vitroplantas de cereza del Perú (*Physalis peruviana* L.). Trop Subtrop Agroecosyst 13:537–542

Mascarenhas LMS, De Santana JRF, Brito AL (2019) Micropropagation of *Physalis peruviana* L. Pesqui Agropecu Trop 49:e55603. https://doi.org/10.1590/1983-40632019v4955603

Mazorra M, Quintana A, Miranda D et al (2003) Análisis sobre el desarrollo y la madurez fisiológica del fruto de la uchuva (*Physalis peruviana* L.) en la zona de Sumapaz (Cundinamarca). Agron Colomb 21:175–189

Miranda D (2005) Criterios para el establecimiento, los sistemas de cultivo, el tutorado y la poda de la uchuva. In: Fischer G, Miranda D, Piedrahíta W, Romero J (eds) Avances en cultivo, poscosecha y exportación de la uchuva (*Physalis peruviana* L.) en Colombia. Universidad Nacional de Colombia, Facultad de Agronomía, Bogotá, pp 29–53

Miranda D, Ulrichs C, Fischer G (2010) Imbibition and percentage of germination of cape gooseberry (*Physalis peruviana* L.) seeds under NaCl stress. Agron Colomb 28:29–35

Moreno N, Álvarez-Herrera J, Balaguera-López H, Fischer G (2009) Propagación asexual de uchuva (*Physalis peruviana* L.) en diferentes sustratos y a distintos niveles de auxina. Agron Colomb 27:341–348

Murashige T, Skoog F (1962) A revised medium for rapid growth and bio assays with tobacco tissue cultures. Physiol Plant 15:473–497. https://doi.org/10.1111/j.1399-3054.1962.tb08052.x

Nuñes AL, Sossmeier S, Gotz AP, Bervian Bispo N (2018) Germination eco-physiology and emergence of *Physalis peruviana* seedlings. J Agric Sci Technol B 8:352–359. https://doi.org/10.17265/2161-6264/2018.06.002

Orozco Orozco L, Trillos González O, Miguel J, Torres Cotes JM (2010) Evaluación de dos métodos de extracción de semillas de uchuva (*Physalis peruviana* L.). Rev Fac Cienc Básicas 6:52–65

Pellizzaro V, Omura MS, Furlan FF et al (2019) Physiological potential of *Physalis peruviana* L. seeds under different temperatures and light wavelengths. Semin Ciênc Agrár 40:1737. https://doi.org/10.5433/1679-0359.2019v40n5p1737

Perea M, Rodríguez N, Fischer G et al (2010) Uchuva: *Physalis peruviana* L. (Solanaceae). In: Perea M, Matallana L, Tirado A (eds) Biotecnología aplicada al mejoramiento de los cultivos de frutas tropicales. Universidad Nacional de Colombia, Bogotá, pp 466–490

Rache Cardenal LY, Pacheco Maldonado JC (2013) Establecimiento de un protocolo de propagación de *Physalis peruviana* L. a partir de yemas axilares adultas. Rev Cienc en Desarollo 4:71–86. https://doi.org/10.19053/01217488.477

Ramar K, Ayyadurai V, Arulprakash T (2014) *In vitro* shoot multiplication and plant regeneration of *Physalis peruviana* L. an important medicinal plant. Int J Curr Microbiol App Sci 3:456–464

Ramírez F, Fischer G, Davenport TL et al (2013) Cape gooseberry (*Physalis peruviana* L.) phenology according to the BBCH phenological scale. Sci Hortic (Amsterdam) 162:39–42. https://doi.org/10.1016/j.scienta.2013.07.033

Rodríguez F, Penoni E, Soares J, Pasqual R (2013) Diferentes concentrações de sais do meio ms e bap na multiplicação *in vitro* de *Physalis peruviana* L. Biosci J 29:77–82

Sandhu A, Singh S, Minhas P, Grewal G (1989) Rhizogenesis of shoot cuttings of raspberry (*Physalis peruviana* L.). Indian J Hortic 46:376–378

Santana G, Angarita A (1997) Regeneración adventicia de somaclones de uchuva (*Physalis peruviana*). Agron Colomb 14:56–65

Singh P, Singh S, Shalitra R et al (2016) *In vitro* regeneration of cape gooseberry (*Physalis peruviana* L.) through nodal segment. Bioscan Int Q J Life Sci 11:41–44

Torres O, Perea M, López A et al (1991) The use of *Physalis peruviana* tissue culture for breeding and seleccion. In: Hawkes J, Lester R, Nee M, Estrada N (eds) Solanaceae 11: taxonomy, chemistry, evolution. Royal Botanic Gardens Kew, Kew, pp 429–432

Verhoeven G (1991) *Physalis peruviana* L. In: VWM V, Coronel R (eds) Plant resources of South-East Asia. No. 2. Edible fruits and nuts. Wageningen (Netherlands). Pudoc, Wageningen, pp 254–256

Yildirim E, Huseyin K, Dursun A (2011) Salt tolerance of *Physalis* during germination and seedling growth. Pak J Bot 43:2673–2676. https://doi.org/10.1590/S0103-90162009000200006

Zeist A, da Silva I, de Lima Filho R et al (2015) Estaquia de *Physalis peruviana* em diferentes bandejas de poliestireno expandido e substratos. Pesq Agrop Gaúcha 21:36–41

Zenkteller M (1972) *In vitro* formation of plants from leaves of several species of the solanaceae family. Biochem Physiol Pflanz 163:509–512

Chapter 8
Fruit Development

Abstract The fruit begins to grow within the calyx after the corolla abscises from the receptacle. The calyx covers the fruit completely during development. It plays a key role in carbohydrate translocation during fruit development. The calyx turns light yellow with brown venation, then light orange, and finally brown. Exocarp (skin) color has been used to develop a maturity index for uchuva fruit. Fruit development depends on environmental conditions and cultivation practices. It is influenced by altitude and temperature in the tropics. Fruit growth and development are characterized by a typical sigmoid growth pattern.

Ovary length ranges from 2.0 to 2.5 mm during fertilization. The corolla abscises, and the fruit begins to grow within the calyx (Figs. 8.1, 8.2 and 8.3) (Valencia 1985; Fischer et al. 2011). The calyx is composed of a green interveinal membrane with dark purple venation during early development (Figs. 8.4 and 8.5). It turns light green with expanding venation as it enlarges (Figs. 8.6 and 8.7). The calyx covers the fruit completely expanding to accommodate the growing fruit during development to form the "Chinese lantern" effect (Figs. 8.5, 8.6, 8.7 and 8.8) (He and Saedler 2005; Fischer and Melgarejo 2014).

The green calyx provides photoassimilates to the growing fruit during the first 20 days of organ development (Figs. 8.5, 8.6, 8.7 and 8.8) (Fischer and Lüdders 1997). It continues to enlarge, turning light yellow with brown venation (Fig. 8.9), and then dark orange as it reaches fruit maturity (Figs. 8.10). The calyx housing the fully mature fruit finally begins to desiccate forming the dried husk (Fig. 8.11).

The fruit develops inside the calyx. Immature fruit are light green, turning dark green as they increase in size (Figs. 8.12, 8.13, 8.14 and 8.15). The enlarging fruit then becomes yellow (Fig. 8.16) and finally orange (Fig. 8.17) as it approaches full size and maturity within the desiccating husk (Fig. 8.18). The interveinal tissues of husks deteriorate to display the vein of non-harvested fruit (Fig. 8.19).

Balaguera-López et al. (2014a) removed or retained the husk from fresh harvested and stored fruits. The resulting endogenous ethylene production was measured in the husk-bearing fruit and in those which the husk was removed. Higher

© The Author(s), under exclusive license to Springer Nature
Switzerland AG 2021
F. Ramírez, T. L. Davenport, *Uchuva (Physalis peruviana L.) Reproductive Biology*, https://doi.org/10.1007/978-3-030-66552-4_8

Fig. 8.1 Corolla abscises (upper photo) and falls to the ground (lower photo). Photos taken in Bogotá, Cundinamarca Colombia

Fig. 8.2 Abscised corolla cut to show anthers and filaments. Photo taken in Bogotá, Cundinamarca Colombia

ethylene production rates in the husk-free fruit were associated with faster ripening and early fruit senescence. Those fruit with intact husks ripened slower due to less ethylene production. Furthermore, such fruits had more firmness, total titratable acidity, and lower weight loss. Angulo (2005) found that optimum conditions for fruit harvesting occur when the calix turns greenish yellow at 18 °C. Fruit harvest is generally based on the color of the calyx; however, fruit color has also been utilized as an indicator of maturity (Fischer et al. 2011).

Exocarp (skin) color has been used to develop a maturity index for uchuva fruit (Galvis et al. 2005). Exocarp color is the most noticeable change occurring to the fruit during maturation and ripening. This color change is from green to yellow. Sbrussi et al. (2014) defined five stages of fruit maturity based on both fruit and calyx color. Green fruit with green calyx is defined as stage **1**, yellow fruit and green calyx is stage **2**, yellow fruit and yellowish green calyx is **3**, yellow fruit with straw-colored calyx is designated stage **4**, and intensely yellow fruit with a desiccated straw-brown husk is stage **5**. Fischer and Martínez (1999) established a calyx color maturity scale from **0** to **6**. Green is **0**, greenish yellow is **1**, yellow-orange is **2**, light orange is **3**, orange is **4**, dark orange is **5** and orange-red is stage **6**. Total soluble solids (TSS) peaked at 17.2% in stage **0** fruits with 9.3 °Brix and then fell to about 13.7% TSS during stage **6**. Icontec (1999) developed a color-based fruit ripeness scale for uchuva. Ripeness color stages were: **0**, green fruit; **1**, light green fruit; **2**, green color at stem end, and orange in the central part of the fruit; **3**, orange-colored fruit with green at stem end; **4**, light orange fruit; **5**, orange fruit; **6**, dark orange fruit. Fruit ripeness is expressed as °Brix associated with each ripeness stage. Thus,

Fig. 8.3 Younger (upper photo) and older (lower photo) calyx with growing ovary and abscising style still attached. Note the deeper purple color in the lower picture. Photos taken in Bogotá, Cundinamarca State, Colombia

Fig. 8.4 Four-lobed ovary enlarging within the calyx after stylar abscission and separation. Lateral view (upper photo). View form bottom (lower photo). Photos taken in Bogotá, Cundinamarca Colombia

Fig. 8.5 Calyx with purple venation completely enclosing the fruit. Photo taken in Bogotá, Cundinamarca Colombia

Fig. 8.6 Young, light-green uchuva calyx. Photo taken in Bogotá, Cundinamarca Colombia

Fig. 8.7 More mature, light-green uchuva calyx. Photo taken in Bogotá, Cundinamarca Colombia

Fig. 8.8 Mature calyx turns light greenish-yellow with purple venation. Photo taken in Bogotá, Cundinamarca Colombia

Fig. 8.9 Uchuva calyx turns light yellow. Photo taken in Bogotá, Cundinamarca State, Colombia

Fig. 8.10 Calyx turns dark orange as it nears fruit maturity. Photo taken in Bogotá, Cundinamarca State, Colombia

Fig. 8.11 Mature calyx begins to desiccate. Photo taken in Bogotá, Cundinamarca State, Colombia

Fig. 8.12 Immature light-green fruit inside surgically opened calyx. Photo taken in Bogotá, Cundinamarca State, Colombia

Fig. 8.13 Young uchuva fruit. Note yellow glandular tissue at the base of the fruit that secrets a protective, waxy coating on surface of the developing fruit (see Chap. 6). Photo taken in Bogotá, Cundinamarca State, Colombia

Fig. 8.14 Continued development of green uchuva fruit within full size calyx. Photo taken in Bogotá, Cundinamarca State, Colombia

Fig. 8.15 Greenish-yellow uchuva fruit growing inside green calyx. Photo taken in Bogotá, Cundinamarca State, Colombia

Fig. 8.16 Yellow, young uchuva fruit. Photo taken in Bogotá, Cundinamarca State, Colombia

Fig. 8.17 Orange uchuva fruit. Photo taken in Bogotá, Cundinamarca State, Colombia

Fig. 8.18 Full sized, mature uchuva fruit with dried husk. Photo taken in Bogotá, Cundinamarca State, Colombia

Fig. 8.19 Husk loses interveinal cells and fruit remains enclosed (upper photo). Detail of husk vasculature (lower photo). Photos taken in Bogotá, Cundinamarca State, Colombia

9.4, 11.4, 13.2, 14.1, 14.5, 14.8, and 15.1 °Brix were associated with color stages 0, 1, 2, 3, 4, 5, and 6, respectively. Moreover, titratable acidity was also associated with each ripeness color stage.

Uchuva fruit development has been characterized by the German Biologische Bundesanstalt, Bundessortenamt und CHemical Industry (BBCH) phenological scale (Ramírez et al. 2013). Fruit development was observed and recorded on a whole-plant basis. Individual fruit ripening was quantified as a percent of full maturity ranging from 10% to 90%.

Mechanical resistance and flesh firmness are important quality attributes displayed after harvest. Ciro Velasquez et al. (2007) and Ciro Velásquez and Osório Saraz (2008) measured the mechanical resistance and flesh firmness of unripe, ripening and ripe uchuva fruits 1, 3, 5, 7, and 9 days after postharvest storage. Low-mass, unidirectional compression force in either longitudinal or transverse direction was then used to measure the mechanical resistance of uchuva fruits at a velocity of 1 mm^{-1s}. Flesh firmness was determined by a 2-mm-diameter puncture needle at 1 mm^{-1s}. Both mechanical resistance and flesh firmness were reduced with increasing postharvest days. Mature fruit were more prone to mechanical damage compared to immature fruits.

Fruit development depends on environmental conditions and cultivation practices (Fischer et al. 2011). Maturity occurred 60–80 days after anthesis when uchuva plants were exposed to constant temperature conditions in a plant growth chamber maintained 16 ± 1 °C and 80% relative humidity (Galvis et al. 2005; Balaguera-López et al. 2014b). In field conditions, fruits reached maturity after 50 days post anthesis in India (Gupta and Roy 1981), after 56–63 days in Germany (Wonneberger 1985) and after 70 days in France (Perón et al. 1989). However, fruit maturity was reached 85–100 days after anthesis when uchuva plants were grown in cooler, high-altitude conditions of Colombia (Fischer and Almanza-Merchán 2012; Fischer and Melgarejo 2014).

Aguilar-Carpio et al. (2018) subjected uchuva plants grown hydroponically to Steiner nutrient solution concentrations of 50, 100 and 150% in greenhouse conditions. Average temperatures ranged from 16 to 21 °C from transplant date to flowering, and 15–18 °C from flowering to fruit ripening. Ripening occurred 110, 115, and 125 days after transplant in the 150, 100 and 50% nutrient solutions, respectively. Faster fruit development, thus, occurred in the stronger nutrient solutions suggesting that nutrients were limited in the two lower concentrations of Steiner nutrient solution.

Uchuva fruit development and maturity are influenced by altitude and temperature in the tropics (Fischer et al. 2011). Fruit maturity required 75 days after floral anthesis at 2690 m average seal level and an average temperature of 12.5 °C in Boyacá State, Colombia (Fischer et al. 2007). Fruits reached maturity 66 days after anthesis at a lower altitude of 2300 m average seal level and higher mean temperature of 17.0 °C (Fischer et al. 2007). Thus, the maturation delay observed with higher altitudes is likely coupled with the associated lower temperatures.

Lima et al. (2012) assessed the physical, chemical, and phytochemical characteristics of uchuva fruits during the harvest period. Fruits were harvested at 120, 150, 180, 210, and 240 days after transplant from seedling germination trays to the field. They were assessed for total mass, skin color, soluble solids (SS) content, phenols, titratable acidity (TA), and total carotenoids and as well as the SS/TA ratio and anti-oxidant activity. Uchuva seeds were germinated using a commercial potting soil and then transplanted to the field when plants acquired two true leaves and were about 20 cm in height. Fruits harvested 210 and 240 days after transplant had the highest mass, soluble solids, phenols, and carotenoids as well as SS/TA ratio.

Uchuva fruit growth and development are characterized by a typical sigmoid growth pattern (Mazorra et al. 2003; Galvis et al. 2005; Fischer et al. 2011). Fischer (2000) and Galvis et al. (2005) reported that fruits exhibit a sigmoidal growth pattern with rapid acceleration in size and weight during the first 10 days of development and continued linear growth until day 60 when the rate of growth slows down until fruit maturity. The calyx becomes fully expanded covering the fruit after only 20–25 days after floral anthesis (Fig. 8.5). At the mature stage, fruits have 12.3 °Brix, pH 3.52, and 1.21 g of citric acid per 100 g fw. Mazorra et al. (2003) described the initial fruit growth phase as slow from fruit set until 20 days after floral anthesis followed by exponential growth occurring from 21 to 60 days. Fruit growth then slows from 61 to 80 days to reach full maturity. Mazorra et al. (2003) also reported slower fruit development time due to cool temperature conditions of 16 and 17.5 °C in San Raimundo and Subia, Cundinamarca State, Colombia, respectively.

Fruit weight gain is highly dependent on climate conditions where plants grow (Angulo 2005). Mean fruit weight was greater in plants growing in open field (5.7 g fw per fruit) conditions than in those growing in greenhouse conditions (5.4 g fw per fruit) (Angulo 2005).

Like tomato, uchuva is climacteric with a high rate of ethylene production during ripening (Gallo 1992; Valdenegro et al. 2012). This endogenously produced ethylene gas triggers a number of metabolic changes during fruit development including ripening and senescence (Balaguera-López et al. 2014b). Ethylene increases the respiratory rate (Valdenegro et al. 2012; Balaguera-López et al. 2014b) as evidenced by elevated carbon dioxide production during the beginning of fruit maturity (Trinchero et al. 1999; Novoa et al. 2002; Valdenegro et al. 2012). Novoa et al. (2006) reported that yellow uchuva fruits had a peak respiratory rate of 33 mg CO_2 kg^{-1} h^{-1} after 12 storage days at 18 °C. The fruit's rise in ethylene biosynthesis and respiration rate are typical of climacteric fruits (Trinchero et al. 1999; Biale 1960). Ethylene production reaches its maximum value during ripening as the latter stage of fruit maturity and decreases with post-ripening senescence (Galvis et al. 2005). Ethylene production rates were 7–24 nmol h^{-1} g^{-1} in the ripe/overripe stages (Trinchero et al. 1999). Alvarado et al. (2004) reported a post-harvest climacteric peak in ethylene production 22–26 days after harvest in Bogotá Colombia. Higher temperatures raised ethylene production. Ethylene is a promoter of fruit softening, causing cell-wall weakening by the activity of hydrolases (Fischer and Bennett 1991). Alvarado et al. (2004) reported that uchuva fruits without a calix had higher respiratory rates than those with an intact calyx after 24 days under storage conditions.

Balaguera-López et al. (2016) studied uchuva fruit maturity during postharvest. Fruits were classified into four stages: S1, 25% yellow and 75% green fruit with green calyx; S2, 50% orange and 50% yellow fruit with yellow-green calyx; S3, 100% orange fruit and 100% yellow calyx, and S4, 100% orange fruit with dry brown calyx. Fruits were stored without calyx at 18 °C and 60% relative humidity for 15 days. Ethylene production, weight loss, color index, TSS, and maturity ratio were higher with increased maturity. Acidity and firmness were lower. Balaguera-López et al. (2017) evaluated the effect of exogenous applied ethylene and its synthesis inhibitor, 1-methylcyclopropene (1-MCP), on the postharvest behavior of Colombia ecotype uchuva fruits, and compared results to those of non-treated, control fruit. Fruit samples were either treated with ethephon (1000 µl l⁻¹), a source of ethylene, or with 1-MCP (1 µl l⁻¹) alone, with 1-MCP applied simultaneously with ethephon, or with 1-MCP applied before ethephon. Fruits treated with 1-MCP delayed ripening and reduced ethylene biosynthesis and respiration rate, total carotenoid content, TSS, skin color development, fruit softening, the loss of titratable acidity, and emission of volatile compounds. Conversely, ethephon treatment increased the physiological changes associated with ripening and prevented the action of the ethylene inhibitor. Gutierrez et al. (2008) applied 1-MCP to samples of immature green, mature green, yellow, and orange uchuva fruits while leaving samples of similar age fruit untreated. Fruits were treated at 25 °C for 20 h and then maintained at 20 °C for 8 days. 1-MCP gas concentrations were 0, 0.5 or 5 µl l⁻¹. Treatment with 1-MCP postponed the start of the ethylene production in immature green and mature green uchuva fruits. It reduced ethylene production in yellow and orange uchuva fruits. Mubarok et al. (2019) studied uchuva fruit treated with 0.5, 1.0, and 2.0 µl l⁻¹ 1-MCP for 6, 12, and 24 h prior to 21 days of postharvest storage. Fruit firmness and titratable acidity (%) were reduced in the fruit after 21 days storage. TSS showed a slight increase. pH increased over time from 5 to 5.67. Balaguera-López et al. (2015) subjected green-yellow uchuva fruits to cool temperature conditions of 2, 6, and 16 °C, with and without application of 1 µl l⁻¹ 1-MCP gas for 35 days. Non-treated fruit subjected to 16 °C took 21 days to ripen. Best uchuva fruit storage results were obtained with the continuous application of 1-MCP at 2 °C. Such fruits stored for 35 days had a reduced respiratory rate, TSS, and color index, and conversely higher total acidity, total carotenoids and firmness.

Total carbohydrate, composed of starch and soluble sugars, increases with uchuva fruit maturity (Galvis et al. 2005; Balaguera-López et al. 2014b). Starch level is high in immature fruits and continues to increase as fruit mature (Fischer 1995). The climacteric rise in ethylene induces synthesis of hydrolytic enzymes that metabolize this starch into reducing sugars, such as glucose and fructose and soluble sugar such as sucrose (TSS), with increasing fruit ripeness (Fischer 1995; Novoa et al. 2006; Fischer and Lüdders 1997). Total sugar content was 63.90 g/kg, which was composed of 31.99 g/kg reducing sugar (Yıldız et al. 2015). Fructose, however, reached a lower concentration than did glucose (Novoa et al. 2006). Glucose and fructose comprise 20% of the dry weight of mature fruits (Galvis et al. 2005). Sucrose content increased in mature harvested fruits during the first 6 days after storage at 12 °C. Thereafter, sucrose concentration deceased until day 24 (Novoa et al. 2006). Fruit physiological maturity is obtained 56 days after anthesis with an increase in TSS in fruit grown in the environmental conditions of Granada, Cundinamarca State, Colombia (Galvis et al. 2005). Thereafter, a decrease in TSS occurs and the pH increases during fruit maturity.

Carbohydrate patterns are similar in fruit and calyx (Fischer and Lüdders 1997). This is because there is a close interrelationship in carbohydrate synthesis, metabolism, and transport between these two organs. While the calyx is green, it acts like a leaf and produces, by photosynthesis, sugar that is exported to the growing fruit where it is converted to starch storage during growth (Balaguera-López et al. 2014b). Fruits without calyx take more time to reach maturity and have higher starch levels than those fruit bearing a calyx.

Organic acids, such as citrate, malate, tartarate, and oxalate decrease during uchuva fruit ripening (Novoa et al. 2006). Citric acid was the most abundant organic acid found in mature fruit, followed by the acids listed above (Novoa et al. 2006). Vitamin C (ascorbic acid) content increases during fruit ripening (Galvis et al. 2005). Its content ranged from 359 to 604 mg 100 g^{-1} fw (Bravo et al. 2015). Valente et al. (2011) reported a mean value of 331 mg 100 g^{-1} of fresh pulp. Furthermore, Vasco et al. (2008) reported ascorbic acid content ranging from 58 to 68 mg 100 g^{-1}.

Da Silva et al. (2016) studied the effect of colored shade nets on the physicochemical characteristics of uchuva fruits in subtropical Brazil. Best quality fruits were produced under white, blue, or black shade nets. In another study, Da Silva et al. (2018) evaluated the effects of shade nets on fruit production and qualitative variables of four species belonging to genus *Physalis*. Uchuva had the best productive and qualitative fruit values under full sunlight or under white, shade-net conditions.

Water- and oxalate-soluble pectic substances increase while those of acid- and alkaline-soluble pectic substances decreased during ripening (Majumder and Mazumdar 2002). There is a 5–six-fold increase in polygalacturonase activity, resulting in degradation of high molecular weight pectin and softening of the pulp; however, pectin methylesterase activity was not clearly related to fruit ripening. Increased polygalacturonase activity level was highly correlated with ethylene biosynthesis although ethylene evolution occurred prior to polygalacturonase synthesis in fruit tissue.

References

Aguilar-Carpio C, Juárez-López P, Campos-Aguilar IH et al (2018) Analysis of growth and yield of cape gooseberry (*Physalis peruviana* L.) grown hydroponically under greenhouse conditions. Rev Chapingo Ser Hortic 24:191–202. https://doi.org/10.5154/r.rchsh.2017.07.024

Alvarado P, Berdugo C, Fischer G (2004) Efecto de un tratamiento de frío (a 1,5°C) y la humedad relativa sobre las características físico-químicas de frutos de uchuva *Physalis peruviana* L. durante el posterior transporte y almacenamiento. Agron Colomb 22:147–159

Angulo R (2005) Crecimiento, desarrollo y producción de la uchuva en condiciones de invernadero y campo abierto. In: Fischer G, Miranda D, Piedrahíta W, Romero J (eds) Avances en cultivo, poscosecha y exportación de la uchuva (*Physalis peruviana* L.) en Colombia. Unibiblos, Bogotá, pp 111–129

Balaguera-López HE, Martínez CCA, Herrera Arévalo A (2014a) Papel del cáliz en el comportamiento poscosecha de frutos de uchuva (*Physalis peruviana* L.) ecotipo Colombia. Rev Colomb Cienc Hortícolas 8:81–191. https://doi.org/10.17584/rcch.2014v8i2.3212

Balaguera-López H, Ramírez Sanabria L, Herrera Arévalo A (2014b) Fisiología y bioquímica del fruto de uchuva (*Physalis peruviana* L.) durante la maduración y poscosecha. In: Carvalho C (ed) Uchuva (*Physalis peruviana* L.) fruta Andina para el mundo. Limencop S.L, Alicante, pp 113–131

Balaguera-López HE, Martínez CA, Aníbal Herrera A (2015) La refrigeración afecta el comportamiento poscosecha de frutos de uchuva (*Physalis peruviana* L.) con cáliz y tratados con 1-metilciclopropeno. Agron Colomb 33:356–364. https://doi.org/10.15446/agron.colomb.v33n3.51896

Balaguera-López HE, Martínez-Cárdenas CA, Herrera-Arévalo A (2016) Effect of the maturity stage on the postharvest behavior of cape gooseberry (*Physalis peruviana* L.) fruits stored at room temperature. Bioagro 28:117–124

Balaguera-López HE, Espinal-Ruiz M, Zacarías L, Herrera AO (2017) Effect of ethylene and 1-methylcyclopropene on the postharvest behavior of cape gooseberry fruits (*Physalis peruviana* L.). Food Sci Technol Int 23:86–96. https://doi.org/10.1177/1082013216658581

Biale JB (1960) The postharvest biochemistry of tropical and subtropical fruits. In: Chichester CO, Mrak EM, Stewart GF (eds) Advances in food research, vol 10. Academic, New York, pp 293–354

Bravo K, Sepulveda-Ortega S, Lara-Guzman O et al (2015) Influence of cultivar and ripening time on bioactive compounds and antioxidant properties in cape gooseberry (*Physalis peruviana* L.). J Sci Food Agric 95:1562–1569. https://doi.org/10.1002/jsfa.6866

Ciro Velásquez H, Osório Saraz J (2008) Avance experimental de la ingeniería de postcosecha de frutas colombianas: resistencia mecánica para frutos de uchuva (*Physalis peruviana* L). DYNA 75:39–46

Ciro Velasquez H, Buitrago Giraldo O, Perez Arango S (2007) Preliminary study of mechanical resistance to fracture and firmness force for uchuva (*Physalis peruviana* L) fruits. Rev Fac Nac Agron 60:3785–3796

Da Silva DF, Pio R, Soares JDR et al (2016) Light spectrum on the quality of fruits of *Physalis* species in subtropical area. Bragantia 75:371–376. https://doi.org/10.1590/1678-4499.463

Da Silva DF, Pio R, Micheli M et al (2018) Productive and qualitative parameters of four *Physalis* species cultivated under colored shade nets. Rev Bras Frutic 40:e-528. https://doi.org/10.1590/0100-29452018528

Fischer G (1995) Effect of root zone temperature and tropical altitude on growth, development and fruit quality of cape gooseberry (*Physalis peruviana* L). Diss. Humboldt-University, Berlin

Fischer G (2000) Crecimiento y desarrollo. In: Flórez V, Fischer G, Sora A (eds) Producción, poscosecha y exportación de la uchuva (*Physalis peruviana* L.). Universidad Nacional de Colombia, Unibiblos, Bogotá, pp 9–26

Fischer G, Almanza-Merchán P (2012) Fisiología del cultivo de la uchuva (*Physalis peruviana* L.). In: II Reuniao Técnica da Cultura da *Physalis*. Lages, pp 32–52

Fischer RL, Bennett AB (1991) Role of cell wall hydrolases in fruit ripening. Annu Rev Plant Physiol Plant Mol Biol 42:675–703. https://doi.org/10.1146/annurev.pp.42.060191.003331

Fischer G, Lüdders P (1997) Developmental changes of carbohydrates in cape gooseberry (*Physalis peruviana* L.) fruits in relation to the calyx and the leaves. Agron Colomb 14:95–107

Fischer G, Martínez O (1999) Calidad y madurez de la uchuva (*Physalis peruviana* L) en relación con la coloración del fruto. Agron Colomb 16:35–39

Fischer G, Melgarejo L (2014) Ecofisiología de la uchuva (*Physalis peruviana* L.). In: Carvalho C (ed) Uchuva (*Physalis peruviana* L.) fruta Andina para el mundo. Limencop S.L, Alicante, pp 29–47

Fischer G, Ebert G, Lüdders P (2007) Production, seeds and carbohydrate contents of cape gooseberry (*Physalis peruviana* L.) fruits grown at two contrasting Colombian altitudes. J Appl Bot Food Qual 81:29–35. https://doi.org/10.17584/rcch.2015v9i2.4177

Fischer G, Herrera A, Almanza PJ (2011) Cape gooseberry (*Physalis peruviana* L.). In: Yahia EM (ed) Postharvest biology and technology of tropical and subtropical fruits. Woodhead Publishing, Oxford, pp 374–397

Gallo F (1992) Postharvest handling, storage and transportation of Colombian fruit. Acta Hortic 310:155–169

Galvis J, Fischer G, Gordillo O (2005) Cosecha y poscosecha de la uchuva. In: Fischer G, Miranda D, Piedrahíta W, Romero J (eds) Avances en cultivo, poscosecha y exportación de la uchuva (*Physalis peruviana* L.) en Colombia. Unibiblos, Bogotá, pp 165–190

Gupta S, Roy S (1981) The floral biology of cape gooseberry (*Physalis peruviana* Linn; Solanaceae, India). Indian J Agric Sci 51:353–355

Gutierrez MS, Trinchero GD, Cerri AM et al (2008) Different responses of goldenberry fruit treated at four maturity stages with the ethylene antagonist 1-methylcyclopropene. Postharvest Biol Technol 48:199–205. https://doi.org/10.1016/j.postharvbio.2007.10.003

He C, Saedler H (2005) Heterotopic expression of MPF2 is the key to the evolution of the Chinese lantern of *Physalis*, a morphological novelty in Solanaceae. Proc Natl Acad Sci U S A 102:5779. https://doi.org/10.1073/pnas.0501877102

Icontec (1999) Norma técnica colombiana uchuva NTC 4580

Lima CSM, Galarça SP, Betemps DL et al (2012) Avaliação física, química e fitoquímica de frutos de *Physalis*, ao longo do período de colheita. Rev Bras Frutic 34:1004–1012. https://doi.org/10.1590/S0100-29452012000400006

Majumder K, Mazumdar BC (2002) Changes of pectic substances in developing fruits of cape-gooseberry (*Physalis peruviana* L.) in relation to the enzyme activity and evolution of ethylene. Sci Hortic (Amsterdam) 96:91–101. https://doi.org/10.1016/S0304-4238(02)00079-1

Mazorra M, Quintana A, Miranda D et al (2003) Análisis sobre el desarrollo y la madurez fisiológica del fruto de la uchuva (*Physalis peruviana* L.) en la zona de Sumapaz (Cundinamarca). Agron Colomb 21:175–189

Mubarok S, Dahlania S, Suwali N (2019) Dataset on the change of postharvest quality of *Physalis peruviana* L. as an effect of ethylene inhibitor. Data Brief 24:038495. https://doi.org/10.1016/j.dib.2019.103849

Novoa R, Bojacá M, Fischer G (2002) Determinación de pérdida de humedad en el fruto de la uchuva (*Physalis peruviana* L.) según el tipo de secado en tres índices de madurez. In: Memorias IV Seminario de Frutales de Clima Frío Moderado, CDTF, Corpoica. CORPOICA, Medellín, pp 298–302

Novoa R, Bojacá M, Galvis J, Fischer G (2006) La madurez del fruto y el secado del cáliz influyen en el comportamiento poscosecha de la uchuva, almacenada a 12 °C (*Physalis peruviana* L.). Agron Colomb 24:77–86

Perón J, Demaure E, Hannetel C (1989) Les possibilities d'introduction et de developpement de solanacees et de cucurbitacees d'origine tropicale en France. Acta Hortic 242:179–186

Ramírez F, Fischer G, Davenport TL et al (2013) Cape gooseberry (*Physalis peruviana* L.) phenology according to the BBCH phenological scale. Sci Hortic (Amsterdam) 162:39–42. https://doi.org/10.1016/j.scienta.2013.07.033

Sbrussi CAG, Zucareli C, Prando AM, Da Silva BV d AB (2014) Maturation stages of fruit development and physiological seed quality in *Physalis peruviana*. Rev Ciênc Agron 45:543–549. https://doi.org/10.1590/S1806-66902014000300015

Trinchero GD, Sozzi GO, Cerri AM et al (1999) Ripening-related changes in ethylene production, respiration rate and cell-wall enzyme activity in goldenberry (*Physalis peruviana* L.), a solanaceous species. Postharvest Biol Technol 16:139–145. https://doi.org/10.1016/S0925-5214(99)00011-3

Valdenegro M, Fuentes L, Herrera R, Moya-León MA (2012) Changes in antioxidant capacity during development and ripening of goldenberry (*Physalis peruviana* L.) fruit and in response to 1-methylcyclopropene treatment. Postharvest Biol Technol 67:110–117. https://doi.org/10.1016/j.postharvbio.2011.12.021

Valencia M (1985) Anatomía del fruto de la uchuva. Acta Agron Colomb 1:63–89

Valente A, Albuquerque TG, Sanches-Silva A, Costa HS (2011) Ascorbic acid content in exotic fruits: a contribution to produce quality data for food composition databases. Food Res Int 44:2237–2242. https://doi.org/10.1016/j.foodres.2011.02.012

Vasco C, Ruales J, Kamal-Eldin A (2008) Total phenolic compounds and antioxidant capacities of major fruits from Ecuador. Food Chem 111:816–823. https://doi.org/10.1016/j.foodchem.2008.04.054

Wonneberger C (1985) Andenbeere – eine alte und neue Kulturpflanze. Gartenpraxis 3:60–61

Yıldız G, İzli N, Ünal H, Uylaşer V (2015) Physical and chemical characteristics of goldenberry fruit (*Physalis peruviana* L.). J Food Sci Technol 52:2320–2327. https://doi.org/10.1007/s13197-014-1280-3

Chapter 9
Fruit Properties and Health Benefits

Abstract Uchuva is a rich source of provitamin A, vitamin B complex (niacin, thiamin, and B12), and vitamin C. The contents of vitamin C, β-carotene, total phenolics, and antioxidant capacity are directly proportional to ripeness stage. Uchuva fruits contain up to 15% soluble solids (mainly sugars) and have a high level of fructose. Numerous medicinal properties have been described in uchuva.

Uchuva fruit (Fig. 9.1) is a source of compounds with potential health benefits for human beings (Table 9.1) (Olivares-Tenorio et al. 2016; Bazalar Pereda et al. 2018; Guiné et al. 2020). It is a rich source of provitamin A, vitamin B complex (niacin, thiamine, and B$_{12}$), and vitamin C (Galvis et al. 2005; Ramadan 2011; Rehm and Espig 1991; Olivares-Tenorio et al. 2016). Vitamin C values ranged from 0.20 to 0.95 g kg^{-1} (Bravo et al. 2015). It is high in β-carotene (up to 2.0 mg per 100 g^{-1} fw) and flavonoids (quercetin, rutin, myricetin, kaempferol, catechin, and epicatechin) (Rehm and Espig 1991; Ramadan 2011; Olivares-Tenorio et al. 2016). The contents of vitamin C, β-carotene, total phenolics, and antioxidant capacity are directly proportional to ripeness stage (Yıldız et al. 2015; Olivares-Tenorio et al. 2016). Citric acid was 1.26% (Sharoba and Ramadan 2011), and titratable acidity ranged from 0.78% to 1.83% in uchuva fruit (Ersoy and Bagci 2011; Sharoba and Ramadan 2011).

Vitamin C content of uchuvas may vary depending on where the fruit is cultivated. For example, fruits of some greenhouse-grown uchuva cultivars grown in the Czech Republic have different vitamin C content, such as Giant, Inka, South Africa, and Golden Berry (Rop et al. 2012). In contrast, no significant differences in vitamin C content were detected among the cultivars, Colombia, Kenya, and South Africa growing in Colombia (Fischer 2000). Vitamin C levels in uchuva fruit grown in different countries and locations are, therefore, not comparable because of the lack of detailed information about cultivation conditions (Olivares-Tenorio et al. 2016). Bazalar Pereda et al. (2018) reported that cultivated fruits grown in the Argentinean Northern Andean region have a higher vitamin C content (32.21 mg ascorbic acid 100 g^{-1} fw) than did those grown in the wild (14.05 mg ascorbic acid 100 g^{-1} fw). Fischer (2000) reported that vitamin C content was not influenced by

F. Ramírez, T. L. Davenport, *Uchuva (Physalis peruviana L.) Reproductive Biology*, https://doi.org/10.1007/978-3-030-66552-4_9

Fig. 9.1 Mature uchuva fruit and husk. Photo taken in Bogotá, Cundinamarca State, Colombia

Table 9.1 Fruit properties and health benefits

Element	Content	Health Benefit	Source
β-Carotene	0.2 mg 100 g^{-1} fw	Important in vision, bone development, cell division / differentiation, and reproduction	Rehm and Espig (1991), Ramadan (2011), Olivares-Tenorio et al. (2016)
Vitamin C	30–46 mg 100 g^{-1} of fresh fruit	Protection against cardiovascular diseases and cancer / improves immune function	
Vitamin B3	26.6 ± 0.9 mg 100 g^{-1} dw	Converts carbohydrates into energy / fat and protein metabolism / regulate nervous system	Vega-Gálvez et al. (2016)
Vitamin B6	24.8 ± 0.2 mg 100 g^{-1} dw		
Vitamin E	1.5 mg g^{-1} fw	Inhibition of the formation of N compounds in the stomach	Barcia et al. (2010), Olivares-Tenorio et al. (2016)
Phenolic compounds	2.5–934.9 mg 100 g^{-1} fw	Inhibition of carcinogenesis	Olivares-Tenorio et al. (2016)
Quercetin	0.1–10.9 mg kg^{-1} fw	Scavenges a wide range of reactive oxygen, nitrogen, and chlorine species	
Rutin	1.7–6.7 mg kg^{-1} fw	Scavenges a wide range of reactive oxygen, nitrogen, and chlorine species	
Myricetin	1.1–1.3 mg kg^{-1} fw	Scavenges a wide range of reactive oxygen, nitrogen, and chlorine species	
Protein	1.66%	Necessary for growth and repair of body tissues / essential constituent of cells	Yıldız et al. (2015)
	0.44 g 100 g^{-1} juice		Ramadan and Mörsel (2007), Puente et al. (2011)
	0.84%		Sharoba and Ramadan (2011)
Lipids	0.18%	Dietary lipids rich in linoleic acid prevent cardiovascular disorders, such as coronary heart disease, atherosclerosis and hypertension	Yıldız et al. (2015)
	0.20 g 100 g^{-1} juice		Ramadan and Mörsel (2007), Puente et al. (2011)
	0.32%		Sharoba and Ramadan (2011)
Carbohydrate	5.56 g 100 g^{-1} juice	Main energy source for body	Ramadan and Mörsel (2007), Puente et al. (2011), Yıldız et al. (2015)
	13.86%		
Total sugar	4.90 g 100 g^{-1} juice		Ramadan and Mörsel (2007)
Total reducing sugars	3.09 g 100 g^{-1} juice		
Total non-reducing sugars	1.81 g 100 g^{-1} juice		

(continued)

Table 9.1 (continued)

Element	Content	Health Benefit	Source
Minerals			
Iron (Fe)	0.1–3.9 mg 100 g^{-1} fw		Leterme et al. (2006), Ramadan and Mörsel (2007), Repo de Carrasco and Encina Zelada (2008), Rodrigues et al. (2009), Torres-Ossandón et al. (2015), Olivares-Tenorio et al. (2016)
Magnesium (Mg)	34.7–120.1 mg 100 g^{-1} fw		
Calcium (Ca)	7.0–37.7 mg 100 g^{-1} fw		
Potassium (K)	55.3–501.9 mg 100 g^{-1} fw		
Phosphorous (P)	34.0–54.9 mg 100 g^{-1} fw		
Sodium (Na)	52.7 mg 100 g^{-1} fw		
Zinc (Zn)	1.5 mg 100 g^{-1} fw		
Copper (Cu)	0.7 mg 100 g^{-1} fw		
Manganese (Mn)	0.7 mg 100 g^{-1} fw		

high altitude in the Andes. Uchuva fruit contains between 20 and 50 mg 100 g^{-1} fw of vitamin C during the edible ripening stages 4–6 (Olivares-Tenorio et al. 2016). Vitamin C is the most abundant water-soluble antioxidant in the human body (Olivares-Tenorio et al. 2016). This vitamin has been associated with protection against cardio vascular diseases, cancer, and to the beneficial effects on immune system functions (Grosso et al. 2013). Furthermore, vitamin C has free radical and reactive oxygen scavenging activity. It also plays a role in promoting collagen formation in the body, inhibition of carcinogenic nitrosamine formation, and participates as substrate in the catecholamine and carnitine biosynthesis as well as protects low density lipoprotein (LDL) cholesterol against oxidation (Grosso et al. 2013).

Vitamin E (α tocopherol) is a fat-soluble compound that is involved in inhibition of nitrogen compound formation in the stomach, protection of selenium against reduction, and inhibition of oxidative damage to polyunsaturated fatty acids in lipid membranes (Olivares-Tenorio et al. 2016). Uchuva fruit collected in Brazil had a tocopherol content of 1.5 g kg^{-1} (Barcia et al. 2010). Total tocopherols were 32.7 g kg^{-1} of extracted lipids in fruit from Egypt (Ramadan and Mörsel 2003). Vega-Gálvez et al. (2016) reported 10.7 ± 0.28 g kg^{-1} of α-tocopherol in the lipid portion of uchuva fruit grown in Colombia. Ramadan and Mörsel (2007) found that uchuva fruit from Egypt contained α tocopherol (28.3 g kg^{-1}), β tocopherol (15.2 g kg^{-1}), γ tocopherol (45.5 g kg^{-1}), and δ tocopherol (1.50 g kg^{-1}) of total lipids. All of these authors show great differences among their findings, which could be due to differences in fruit and / or cultivars grown in different conditions or to differences in biochemical extraction methods. More refined methodologies may need to be considered for vitamin E extraction of uchuva.

The B Vitamins have also been examined in uchuva fruit. Olivares-Tenorio et al. 2016 reported that vitamin B6 (pyridoxine) and B3 (niacin) convert carbohydrates into energy for body functions, aid in metabolism of fats and proteins, and also regulate the nervous system. Vitamins B3 and B6 in uchuva fruit pulp are 26.6 ± 0.9 mg 100 g^{-1} dw and 24.8 ± 0.2 mg 100 g^{-1} dw, respectively (Vega-Gálvez et al. 2016).

Phenolic compounds are secondary metabolites that occur in plants. They have been linked to inhibition of cancer development through their antioxidant capacity (Olivares-Tenorio et al. (2016). The main phenolic acids identified in uchuva were caffeic, chlorogenic, ferulic, p-coumaric, and gallic acids (Olivares-Tenorio et al. 2016). Total phenolic compounds in uchuva range from 2.5 to 934.9 mg 100 g^{-1} fw, usually expressed as gallic acid equivalents (GAE) (Olivares-Tenorio et al. 2016). This large variation is explained by cultivar differences (Rop et al. 2012). Valdenegro et al. (2013) reported that uchuva fruit has a total phenolic content of 2.46 g GAE 100 g^{-1} dw, and Bravo et al. (2015) found that the total phenolic compounds ranged from 0.06 to 0.74 g GAE kg^{-1} fw. Antioxidant activity in fruit collected form cultivated uchuva plants in Argentina was double that of fruit collected form wild type uchuva plants (Bazalar Pereda et al. 2018).

Flavonoids scavenge a wide range of reactive oxygen, nitrogen, and chlorine species, such as superoxide, hydroxyl radical, peroxyl radicals, hypochlorous acid, and peroxynitrous acid (Olivares-Tenorio et al. 2016). They can also chelate metal ions, often decreasing metal ion pro-oxidant activity. This activity contributes to the prevention of major age-related diseases (Del Rio et al. 2013). The main flavonoids identified and quantified in uchuva were quercetin (0.1–10.9 mg kg^{-1}), rutin (1.7–6.7 mg kg^{-1}), myricetin (1.1–1.3 mg kg^{-1}), epicatechin (0.2–0.6 mg kg^{-1}), and catechin (3.8–6.7 mg kg^{-1}) (Olivares-Tenorio et al. 2016). Extraction methods have been inconsistent giving rise to a wide range of results.

Carotenoids that produce the yellow-orange color of uchuva fruit, are convertible to vitamin A (approx. 10%) by enzymatic cleavage in the human body (Olivares-Tenorio et al. 2016). They participate in cancer prevention due to their pro-vitamin A activity. This is important for the normal maintenance of epithelial cellular differentiation (Fiedor and Burda 2014). β-carotene is a quencher of singlet oxygen inhibiting lipid peroxidation. It is the most important carotenoid in uchuva and gives the yellow-orange color to fruit (Fischer 2000). β-carotene is important in vision, bone development, cell division and differentiation, and reproduction (Bazalar Pereda et al. 2018). Large variation in its concentration has been reported among studies ranging from 0.2 to 1074,7 mg β-carotene 100 g^{-1} fw (Olivares-Tenorio et al. 2016). These differences can partly be attributed to cultivar differences. Several uchuva studies have revealed an increase in β-carotene with ripeness (Severo et al. 2010; Bravo et al. 2015). Wild uchuva fruit grown in the Andean region of Argentina had quantitatively higher β-carotene content (1.99 mg 100 g^{-1} fw) than did cultivated ones (1.24 mg 100 g^{-1} fw) (Bazalar Pereda et al. 2018). Uchuva fruits were determined to have different carotenoid profiles during ripening states and in different fruit fractions such as peel, pulp, and calyx of ripe fruits in Egypt (Etzbach et al. 2018). The carotenoid profile of fruits is dominated by lutein (51%). The profile in ripe fruits consisted of β-carotene (55%) and several carotenoid fatty acid esters, particularly lutein esters esterified with myristic and palmitic acid as monoesters or diesters. Overripe fruits had a significantly lower level of total carotenoids (31%) due to degradation. The peel of ripe fruit had 2.8 times higher total carotenoid content (332.00 μg g^{-1} fw) than did fruit pulp.

Fruit dry matter, water-soluble dry matter, ash, protein, oil, and carbohydrate components were found to be 18.67%, 14.17%, 2.98%, 1.66%, 0.18% and 13.86%, respectively, in uchuva fruits harvested in Turkey (Yıldız et al. 2015). Similarly, Sharoba and Ramadan (2011) reported that uchuva fruit obtained in Egypt had 21% dry matter, 16% water-soluble dry matter, 1.08% ash, 0.84% protein, and 0.32% oil. Total lipids within the berry were 2.0% in Egipt (Ramadan and Mörsel 2019). Triacylglycerols were the main lipid class and comprised nearly 81% of total neutral lipids in the whole berry oil. The oils extracted from fruits of uchuva contain 15 fatty acids i.e. linoleic, oleic, palmitic, and stearic acids, which constitute 95% of the total fatty acids (Puente et al. 2011). Moreover, linoleic acid derivatives are structural components within plasma membranes and are precursors and metabolic regulators of some compounds. Stearic acid (~2.5%) and palmitic acid (9%) are saturated fatty acids mainly found in oils extracted from uchuva fruit (Ramadan and Mörsel 2003). Other lipids found in uchuva were γ-linolenic acid, α-linolenic and dihomo-γ-linolenic acids. Linolenic and linoleic acids are considered essential fatty acids (EFA) since they are necessary for good health. EFAs are important in the synthesis of many cellular structures and several biologically important compounds (Latham 2002).

Uchuva fruits contain up to 15% soluble solids (mainly sugars) and have a high level of fructose. They are rich in dietary fiber, and pectin, which are important in large intestine regularity (Ramadan 2011).

Bazalar Pereda et al. (2018) reported that cultivated uchuva fruit had higher magnesium (48.70 mg 100 g^{-1} fw) and copper (0.35 mg 100 g^{-1} fw) content than did wild uchuva plants (35.96 Mg mg 100 g^{-1} fw and 0.26 mg Cu 100 g^{-1} fw). Potassium is mainly an intracellular cation, taking part in the cellular uptake of molecules against electrochemical and concentration gradients (Olivares-Tenorio et al. 2016). Potassium also takes part in the electrophysiology of nerves and muscle and in acid-base regulation. Table 9.1 summarizes the mineral content of uchuva after Leterme et al. (2006); Ramadan and Mörsel (2007); Repo de Carrasco and Encina Zelada (2008); Rodrigues et al. (2009); Torres-Ossandón et al. (2015); Olivares-Tenorio et al. (2016).

Withanolides are steroidal lactones produced mainly by solanaceous plant species (Puente et al. 2011). They have a broad spectrum of biological properties and pharmacological activities that include insect-repellent, insect-antifeedant, antibacterial, hepatoprotective, immunomodulatory, antitumor, anti-inflammatory, and cytotoxic activity, as well as protection against induced hepatoxicity (Ramadan 2011).

Numerous medicinal properties have been recognized for *P. peruviana* such as antispasmodic, antiseptic, sedative, diuretic, analgesic, helping to fortify the optic nerve, throat trouble relief, and elimination of intestinal parasites and pathogenic protozoans, such as amoebae (Puente et al. 2011).

Uchuva fruits have been used in traditional medicine of Peru to treat cancer, hepatitis, asthma, malaria and dermatitis. Such uses, however, have not been scientifically authenticated (Zavala et al. 2006).

In medical trials, ethanolic extracts of fresh uchuva fruits and acetone-dehydrated fruit extracts improved culture viability of astrocytic cell lines by reducing the formation of reactive oxygen species to preserve mitochondrial membrane potential under oxidative stress with rotenone (Areiza-Mazo et al. 2018).

References

Areiza-Mazo N, Robles J, Zamudio-Rodriguez JA et al (2018) Extracts of *Physalis peruviana* protect astrocytic cells under oxidative stress with rotenone. Front Chem 6:276. https://doi.org/10.3389/fchem.2018.00276

Barcia M, Jacques A, Pertuzatti P, Zambiazi R (2010) Determination by HPLC of ascorbic acid and tocopherols in fruits. Semina Cienc Agrar 31:381–390

Bazalar Pereda M, Nazareno M, Viturro C (2018) Nutritional and antioxidant properties of *Physalis peruviana* L. fruits from the Argentinean northern Andean region. Plant Foods Hum Nutr 74:1–8

Bravo K, Sepulveda-Ortega S, Lara-Guzman O et al (2015) Influence of cultivar and ripening time on bioactive compounds and antioxidant properties in cape gooseberry (*Physalis peruviana* L.). J Sci Food Agric 95:1562–1569. https://doi.org/10.1002/jsfa.6866

Del Rio D, Rodriguez-Mateos A, Spencer JPE et al (2013) Dietary (poly)phenolics in human health: structures, bioavailability, and evidence of protective effects against chronic diseases. Antioxid Redox Signal 18:1818–1892. https://doi.org/10.1089/ars.2012.4581

Ersoy N, Bagci Y (2011) Some physico-chemical properties and antioxidant activities of goldenberry (*Physalis peruviana* L.), pepino (*Solanum muricatum* ait.) and passiflora (*Passiflora edulis* Sims) tropical fruits. Univ Selcuk J Agric Food Sci 25:67–72

Etzbach L, Pfeiffer A, Weber F, Schieber A (2018) Characterization of carotenoid profiles in goldenberry (*Physalis peruviana* L.) fruits at various ripening stages and in different plant tissues by HPLC-DAD-APCI-MS^n. Food Chem 245:508–517. https://doi.org/10.1016/j.foodchem.2017.10.120

Fiedor J, Burda K (2014) Potential role of carotenoids as antioxidants in human health and disease. Nutrients 6:466–488. https://doi.org/10.3390/nu6020466

Fischer G (2000) Crecimiento y desarrollo. In: Flórez V, Fischer G, Sora A (eds) Producción, poscosecha y exportación de la uchuva (*Physalis peruviana* L.). Universidad Nacional de Colombia, Unibiblos, Bogotá, pp 9–26

Galvis J, Fischer G, Gordillo O (2005) Cosecha y poscosecha de la uchuva. In: Fischer G, Miranda D, Piedrahíta W, Romero J (eds) Avances en cultivo, poscosecha y exportación de la uchuva (*Physalis peruviana* L.) en Colombia. Unibiblos, Bogotá, pp 165–190

Grosso G, Bei R, Mistretta A et al (2013) Effects of vitamin C on health: a review of evidence. Front Biosci 18:1017–1029. https://doi.org/10.2741/4160

Guiné RPF, Gonçalves FJA, Oliveira SF, Correia PMR (2020) Evaluation of phenolic compounds, antioxidant activity and bioaccessibility in *Physalis peruviana* L. Int J Fruit Sci 20:S470. https://doi.org/10.1080/15538362.2020.1741056

Latham M (2002) Macronutrientes: carbohidratos, proteínas y grasas. Nutr Humana en el Mundo en Desarro 29:99–107

Leterme P, Buldgen A, Estrada F, Londoño AM (2006) Mineral content of tropical fruits and unconventional foods of the Andes and the rain forest of Colombia. Food Chem 95:644–652. https://doi.org/10.1016/j.foodchem.2005.02.003

Olivares-Tenorio ML, Dekker M, Verkerk R, van Boekel MAJS (2016) Health-promoting compounds in cape gooseberry (*Physalis peruviana* L.): review from a supply chain perspective. Trends Food Sci Technol 57:83–92. https://doi.org/10.1016/j.tifs.2016.09.009

Puente LA, Pinto-Muñoz CA, Castro ES, Cortés M (2011) *Physalis peruviana* Linnaeus, the multiple properties of a highly functional fruit: a review. Food Res Int 44:1733–1740. https://doi.org/10.1016/j.foodres.2010.09.034

Ramadan MF (2011) Bioactive phytochemicals, nutritional value, and functional properties of cape gooseberry (*Physalis peruviana*): an overview. Food Res Int 44:1830–1836. https://doi.org/10.1016/j.foodres.2010.12.042

Ramadan MF, Mörsel JT (2003) Oil goldenberry (*Physalis peruviana* L.). J Agric Food Chem 51:969–974. https://doi.org/10.1021/jf020778z

Ramadan MF, Mörsel JT (2007) Impact of enzymatic treatment on chemical composition, physicochemical properties and radical scavenging activity of goldenberry (*Physalis peruviana* L.) juice. J Sci Food Agric 87:452–460. https://doi.org/10.1002/jsfa.2728

Ramadan MF, Mörsel J-T (2019) Goldenberry (*Physalis peruviana*) oil. In: Ramadan MF (ed) Fruit oils: chemistry and functionality. Springer International Publishing, Cham, pp 397–404

Rehm S, Espig G (1991) The cultivated plants of the tropics and subtropics. Verlag Josef Margraf, Weikersheim

Repo de Carrasco R, Encina Zelada CR (2008) Determinación de la capacidad antioxidante y compuestos bioactivos de frutas nativas peruanas. Rev Soc Quím Perú 74:108–124

Rodrigues E, Rockenbach II, Cataneo C et al (2009) Minerals and essential fatty acids of the exotic fruit *Physalis peruviana* L. Ciênc Tecnol Aliment 29:642–645. https://doi.org/10.1590/S0101-20612009000300029

Rop O, Mlcek J, Jurikova T, Valsikova M (2012) Bioactive content and antioxidant capacity of cape gooseberry fruit. Cent Eur J Biol 7:672–679. https://doi.org/10.2478/s11535-012-0063-y

Severo J, Lima C, Coelho M et al (2010) Atividade antioxidante e fitoquímicos em frutos de Physalis (*Physalis peruviana*, L.) durante o amadureçimento e o armazenamento. Rev Bras Agrociências 16:77–82

Sharoba AM, Ramadan MF (2011) Rheological behavior and physicochemical characteristics of goldenberry (*Physalis peruviana*) juice as affected by enzymatic treatment. J Food Process Preserv 35:201–219. https://doi.org/10.1111/j.1745-4549.2009.00471.x

Torres-Ossandón MJ, López J, Vega-Gálvez A et al (2015) Impact of high hydrostatic pressure on physicochemical characteristics, nutritional content and functional properties of cape gooseberry pulp (*Physalis peruviana* L.). J Food Process Preserv 39:2844–2855. https://doi.org/10.1111/jfpp.12535

Valdenegro M, Almonacid S, Henríquez C et al (2013) The effects of drying processes on organoleptic characteristics and the health quality of food ingredients obtained from goldenberry fruits (*Physalis peruviana*). Open Acess Sci Rep 2:642. https://doi.org/10.4172/scientificreports

Vega-Gálvez A, Díaz R, López J et al (2016) Assessment of quality parameters and microbial characteristics of cape gooseberry pulp (*Physalis peruviana* L.) subjected to high hydrostatic pressure treatment. Food Bioprod Process 97:30–40. https://doi.org/10.1016/J.FBP.2015.09.008

Yıldız G, İzli N, Ünal H, Uylaşer V (2015) Physical and chemical characteristics of goldenberry fruit (*Physalis peruviana* L.). J Food Sci Technol 52:2320–2327. https://doi.org/10.1007/s13197-014-1280-3

Zavala D, Mauricio Q, Pelayo A et al (2006) Citotoxic effect of *Physalis peruviana* (capuli) in colon cancer and chronic myeloid leukemia. An Fac Med 67:283–289

Chapter 10
Breeding and Hybridization

Abstract In the year 2000, there were 12 *Physalis* germplasm repositories in Latin America and the Caribbean. Colombia's National Germplasm Repository, a division of Colombia's Corporation of Agricultural Research (AGROSAVIA) houses 98 uchuva accessions. Other Colombian Universities house up to 222 uchuva accessions. The Experimental Station, Santa Catalina, in Ecuador (Estación Experimental Santa Catalina, DENAREF, INIAP) houses a germplasm repository with 23 traditional varieties of *P. peruviana* from Ecuador, one from Colombia, and one from Bolivia.

Germplasm repositories are important for collection, breeding, assessment of genetic diversity, and general improvement of plant accessions. They have played an important role in the study of tropical solanaceous species, such as tree tomato, lulo, uchuva, and cocona (Valencia et al. 2010; Ramírez et al. 2018; Ramírez and Kallarackal 2019; Ramírez 2020, 2021). Knudsen (2000) reported that by the year 2000, there were 12 *Physalis* spp. germplasm repositories in Latin America and the Caribbean. Out of the 12, there were 4 that belonged to *Physalis peruviana*. These collections included 75 accessions of which, 39 were housed in the germplasm repository, La Selva in Colombia, 23 in the Estación Experimental Santa Catalina, DENAREF, INIAP – Ecuador, 1 in CIRAD-FLHOR, Guadeloupe island station, France, and 12 in the Universidad Nacional San Antonio Abad del Cusco, Perú. Valencia et al. (2010) and Chacón et al. (2016) reported that Colombia's National Germplasm Repository, a division of Colombia's Corporation of Agricultural Research (AGROSAVIA) housed a collection of 57 uchuva accessions (48 of these are native to Colombia and nine are from other countries). AGROSAVIA houses 98 accessions in two research stations namely Rionegro in Antioquia State and Tibaitatá in Cundinamarca State, Colombia (Ligarreto et al. 2005; Fischer et al. 2011). Furthermore, Universidad Nacional de Colombia, Bogotá, houses 54 uchuva accessions and Universidad de Nariño, Pasto houses 50 accessions (Lagos et al. 2003). Universidad Nacional de Colombia, Palmira, has a collection of 222 uchuva accessions (Ligarreto et al. 2005; Betancourt et al. 2008). The accessions housed in Palmira were collected in six states, Nariño, Valle del Cauca, Cauca, Caldas,

F. Ramírez, T. L. Davenport, *Uchuva (Physalis peruviana L.) Reproductive Biology*, https://doi.org/10.1007/978-3-030-66552-4_10

Cundinamarca, and Quindío (Betancourt et al. 2008). Although Colombia has several uchuva collections, this country lacks a reliable selection system and plant material origin has not been determined in some cases (Ligarreto et al. 2005). Some uchuva collections have been partly evaluated for resistance to *Fusarium oxysporum* for improvement (Chacón et al. 2016). Universidad Nacional de Colombia and AGROSAVIA have studied uchuva ecotypes from five states in Colombia, namely Antioquia, Cundinamarca, Boyacá, Santander, and Norte de Santander (Fig. 10.1) (Herrera et al. 2011, 2012; Enciso-Rodríguez et al. 2013; Berdugo Cely et al. 2015; Garzón-Martínez et al. 2012). These investigators reported that cultivated uchuva accessions have a higher yield and fruit weight, whereas wild accessions have a higher number of fruits per plant, absence of fruit cracking, and high concentration of total soluble solids. Furthermore, wild accessions exhibit tolerance to *F. oxysporum* in Nariño state, Colombia. Accession features of uchuva, such as fruit quality and yield, are key aspects in hybridization and breeding programs (Prohens and Nuez 1994; Leiva-Brondo et al. 2001). Uchuva hybrids can increase yield without affecting fruit quality (Leiva-Brondo et al. 2001).

Morphological and genetic data bases have been established for *Physalis* in Colombia (Ligarreto et al. 2005). This information has been key for establishing plant collections. The Palmira uchuva collection represents native and cultivated *P. peruviana* plants occurring in Nariño, Valle del Cauca, Caldas, and Cundinamarca states in Colombia. The AGROSAVIA *Physalis* collection has introductions from Guatemala, the botanic garden of Nijmegen in the Netherlands, and from Antioquia, Caldas, Cundinamarca, and Nariño states in Colombia (CORPOICA 2004).

The Experimental Station, Santa Catalina, in Ecuador (Estación Experimental Santa Catalina, DENAREF, INIAP) houses a germplasm repository with 23 traditional varieties of *P. peruviana* from Ecuador, one from Colombia and one from Bolivia (Knudsen 2000). Uchuva ecotypes have been selected for cultivation by Cusco and Cajamarca Universities in Peru (Tapia and Fries 2007). Six uchuva ecotypes occur in the northern region of Peru, namely Urquiaco and Agocucho, Cajabamba and Era in Cajamarca State and Huancayo 1 and Huancayo 2 in Junín State (Sánchez 2006).

Hybridization among *Physalis* species has been difficult to obtain. Menzel (1951) reported that hybridization has been unsuccessful among *P. peruviana* and other related species. *P. ixocarpa* Rydb was crossed with *P. peruviana* obtaining mature seeds; however, F1 and F2 progenies from both were uniformly identical with the seed parents, showing no evidence of hybridization. Fruits were not obtained when *P. ixocarpa* was the seed parent or when *ixocarpa* pollen was used on flowers of perennial species. Few investigations have reported successful hybrids. Azeez and Faluyi (2018) attempted interspecific reciprocal crosses among four *Physalis* species; *P. angulata, P. micrantha, P. pubescens,* and *P. peruviana.* They hand-transferred desired pollen grains to specific parents. The only viable hybrid F1 cross was *P. angulata* x *P. pubescens.* This hybrid produced few flower buds, and no mature fruits were harvested. Hybrid plant pollen diameter was 25.40–30.72 μm, thus, falling within the range of those of the two progenitors. The F1 hybrid pollen cells suffered meiotic irregularities. The hybrid was triploid, having a chromosome

Fig. 10.1 Colombian ecotype near Tuta, Boyacá State, Colombia

number of 23 n. Pal et al. (1993) reported that hybrids were successfully obtained from crosses between *P. minima* L. and *P. peruviana* by culturing the excised embryos *in vitro* after 20–25 days post pollination on MS medium supplemented with kinetin (0.2 mg l^{-1}) and α-naphthaleneacetic acid (0.1 mg l^{-1}). The crossed hybrid fruits had the same color as that of wild male parent but with a varied intensity. The regenerated plantlets were transferred to pots and watered with nutrient solution. Overall, 64% of transferred plantlets established well. Ganapathi et al. (1991) recorded unsuccessful hybridization between *P. pubescens* and *P. peruviana*, and between *P. pubescens* and *P. angulata* including their reciprocal crosses.

Leiva-Brondo et al. (2001) studied vegetative characters such as yield, fruit weight, fruit shape, soluble solids content, titratable acidity, and ascorbic acid content in three uchuva hybrids, and their parents under field and greenhouse conditions. The highest yields were obtained with hybrids growing in a glasshouse in Valencia, Spain. Interaction dominance vs environment was important for expression of yield. Yield was higher in plants grown in a greenhouse than those grown in the field. Broad-sense heritability for all characters was high to medium (0.48–0.91) indicating that a high response to selection should be expected (Leiva-Brondo et al. 2001).

Berdugo Cely et al. (2015) morphologically characterized four genotypes of *P. peruviana* and one of *P. floridana* using 34 and 20 qualitative and quantitative morphological variables, respectively. These authors also used 328 Conserved Orthologous Sequence (COSII) molecular markers and 154 Immunity Related Gene (IRGs) markers. This study focused on the F1 generation and the molecular characterization of the population. Quantitative variables differentiated the species *P. peruviana* and *P. floridana*. F1 viability was 50% between *P. floridana* and *P. peruviana*, but only when the former was the pollen receptor.

References

Azeez S, Faluyi J (2018) Hybridization in four Nigerian *Physalis* (Linn.) species. Not Sci Biol 10:205–210

Berdugo Cely J, Rodríguez F, González Almario C, Barrero Meneses L (2015) Variabilidad genética de parentales y poblaciones F1 inter e intraespecíficas de *Physalis peruviana* L. y *P. floridana* Rydb. Rev Bras Frutic 37:179–192

Betancourt MLB, Piedrahíta KE, Terranova AMP et al (2008) Establecimiento de una colección de trabajo de uchuva del suroccidente colombiano. Acta Agron 57:95–99

Chacón M, Sánchez Y, Barrero L (2016) Genetic structure of a Colombian cape gooseberry (*Physalis peruviana* L.) collection by means of microsatellite markers. Agron Colomb 34:5–16. https://doi.org/10.15446/agron.colomb.v34n1.52960

CORPOICA (2004) Inventario de bancos de germoplasma. Corporación Colombiana de Investigación Agropecuaria – Corpoica. Rionegro, Antioquia

Enciso-Rodríguez FE, González C, Rodríguez EA et al (2013) Identification of immunity related genes to study the *Physalis peruviana – Fusarium oxysporum* Pathosystem. PLoS One 8:e68500. https://doi.org/10.1371/journal.pone.0068500

Fischer G, Herrera A, Almanza PJ (2011) Cape gooseberry (*Physalis peruviana* L.). In: Yahia EM (ed) Postharvest biology and technology of tropical and subtropical fruits. Woodhead Publishing, Oxford, pp 374–397

Ganapathi A, Sudhakaran S, Kulothungan S (1991) The diploid taxon in Indian natural populations of *Physalis* L. and its taxonomic significance. Cytologia (Tokyo) 56:283–288

Garzón-Martínez GA, Zhu Z, Landsman D et al (2012) The *Physalis peruviana* leaf transcriptome: assembly, annotation and gene model prediction. BMC Genomics 13:151. https://doi.org/10.1186/1471-2164-13-151

Herrera A, Fischer G, Chacón MM (2012) Agronomical evaluation of cape gooseberries (*Physalis peruviana* L.) from central and north-eastern Colombia. Agron Colomb 30:15–24

Herrera A, Herrera M, Ortiz A et al (2011) Comportamiento en producción y calidad de 54 accesiones de uchuva (*Physalis peruviana* L.) provenientes del nor-oriente colombiano. Agron Colomb 29:189–196

Knudsen H (2000) Directorio de colecciones de germoplasma en América Latina y el Caribe. International Plant Genetic Resources Institute (IPGRI), Rome

Lagos T, Criollo H, Ibarra A, Hejeile H (2003) Caracterización morfológica de la colección Nariño de uvilla o uchuva *Physalis peruviana*. Fitotec Colomb 3:1–9

Leiva-Brondo M, Prohens J, Nuez F (2001) Genetic analyses indicate superiority of performance of cape gooseberry (*Physalis peruviana* L.) Hybrids. J New Seeds 3:71–84. https://doi.org/10.1300/J153v03n03_04

Ligarreto G, Lobo M, Correa M (2005) Recursos genéticos del género *Physalis* en Colombia. In: Fisher G, Miranda D, Piedrahita W, Romero J (eds) Avances en cultivo, poscosecha y exportación de la uchuva (*Physalis peruviana* L.) en Colombia. Universidad Nacional de Colombia, Unibiblos, Bogotá, pp 9–26

Menzel M (1951) The cytotaxonomy and genetics of *Physalis*. Proc Am Philos Soc 95:132–183

Pal B, Bindra A, Raman H (1993) *In vitro* plantlet regeneration from embryos of inter-specific hybrids of cape gooseberry (*Physalis peruviana* L.). Indian J Plant Genet Resour 6:40

Prohens J, Nuez F (1994) Aspectos productivos de la introducción de nuevos cultivares de alquequenje (*Physalis peruviana* L.) en España. Acta Hortic 12:228–233

Ramírez F, Kallarackal J (2019) Tree tomato (*Solanum betaceum* Cav.) reproductive physiology: a review. Sci Hortic (Amsterdam) 248:206–215

Ramírez F, Kallarackal J, Davenport TL (2018) Lulo (*Solanum quitoense* Lam.) reproductive physiology: a review. Sci Hortic (Amsterdam) 238:163–176. https://doi.org/10.1016/j.scienta.2018.04.046

Ramírez F (2020) Cocona (*Solanum sessiliflorum* Dunal) reproductive physiology: a review. Genet Resour Crop Evol 67:293–311

Ramírez F (2021) Notes about Lulo (*Solanum quitoense* Lam.): an important South American underutilized plant. Genet Resour Crop Evol 68:93–100. https://doi.org/10.1007/s10722-020-01059-3

Sánchez H (2006) Evaluación agronómica de seis ecotipos de "tomatillo" (*Physalis peruviana*) para su adaptación en tres pisos ecológicos de la cuenca alta del Llaucano. Universidad Nacional de Cajamarca

Tapia M, Fries M (2007) Guía de campo de los cultivos andinos. Organización de las Naciones Unidas para la Agricultura y la Alimentación (FAO) y Asociación Nacional de Productores Ecológicos del Perú

Valencia R, Lobo M, Ligarreto G (2010) Estado del arte de los recursos genéticos vegetales en Colombia: Sistema de Bancos de Germoplasma. Corpoica Cienc Tecnol Agropecu 11:85–94

Chapter 11
Genetic Diversity

Abstract Molecular techniques have been applied to study uchuva genetic diversity. These include Random Amplified Microsatellites (RAMs), Simple Sequence Repeat Markers (SSR), Single Nucleotide Polymorphism (SNP) markers, and Random Amplified Polymorphic DNA (RAPD) primers. Other techniques utilized protein denaturation and separation using polyacrylamide gel electrophoresis (SDS-PAGE). Uchuva is a diploid plant with 24 chromosomes ($2n = 24$). Some uchuva ecotypes have ploidy variations between diploid ($2n = 24$) and tetraploid ($2n = 48$).

Genetic diversity of uchuva was poorly studied before 2001. Since then, a number of investigations have focused on genetic diversity at the molecular level. Morillo et al. (2001) found 66 polymorphic loci with bands of 500–3000 base pairs using the Random Amplified Microsatellites (RAMs) technique. The value of heterozygosity for 18 introductions was 0.44, and the largest genetic distance between the Ecotypes, Colombia and Peru was 1.42. Bonilla et al. (2008) assessed the genetic diversity of 43 accessions of uchuva from Nariño, Valle del Cauca, Cauca, Caldas, Quindío, and Cundinamarca states in Colombia. Their results showed a high genetic diversity among accessions from Valle del Cauca and low genetic variation among accessions from Nariño, Quindío, and Cauca states. Muñoz et al. (2008) identified high genetic variability and distant groups within *P. peruviana* using RAMs. Morillo-Coronado et al. (2018) used seven RAM markers to characterize samples collected from 15 wild-type plants growing in Paipa, Combita, Tuta, Sotaquirá, Santa Rosa de Viterbo, and Tunja, in Boyacá State, Colombia. This population segregated into three groups according to the local origins of the plants, based on the Nei-Li coefficient at a similarity level of 0.65. The average heterozygosity was 0.27, which indicates close interrelationship among the 15 sampled plants. Simbaqueba et al. (2011) established the first collection of Simple Sequence Repeat Markers (SSR) derived from *P. peruviana* commercial Colombian genotype. They found that 1236 out of 1520 SSR loci were composed of imperfect repeats, thus, increasing the likelihood of genetic diversity of *P. peruviana*. Simbaqueba et al. (2011) identified a total of 932 imperfect and 201 perfect SSR loci in untranslated regions (UTRs) of the DNA and

F. Ramírez, T. L. Davenport, *Uchuva (Physalis peruviana L.) Reproductive Biology*, https://doi.org/10.1007/978-3-030-66552-4_11

304 imperfect and 83 perfect SSR loci in coding regions from the assembled *P. peruviana* leaf transcriptome. Furthermore, UTR and SSR loci were employed to develop 162 primers for amplification. The efficiency of these primers was tested through PCR in a panel of seven *P. peruviana* accessions that included Colombia, Kenya, and Ecuador ecotypes and one closely related species, *Physalis floridana*. Having obtained a PCR amplification rate of 83% the investigators found the genetic diversity among ecotypes to be low with a polymorphic rate of 22%.

Garzón-Martínez et al. (2015) studied the genetic diversity and population structure of *P. peruviana*. They characterized 47 uchuva accessions and 13 accessions of related species consisting of 222 individuals from the AGROSAVIA germplasm collection in Colombia. To do this, they used Conserved Orthologous Sequence (COSII) and Immunity Related Gene (IRGs) protocols. Six hundred forty two Single Nucleotide Polymorphism (SNP) markers were identified and used within the genetic diversity analysis. Enciso-Rodríguez et al. (2013) recognized 74 IRGs from uchuva, then selected 17 markers that were sequenced in a small subset of uchuva and related species allowing the identification of one candidate SNP gene associated with the resistance response against the fungal pathogen *Fusarium oxysporum*. Osorio-Guarín et al. (2016) analyzed the genetic diversity of disease resistance gene from a collection of 100 Colombian uchuva accessions using 5000 SNP markers and concluded that gene expression was mainly associated with the degree of commercial improvement or wildness and not to its geographic origin.

Trevisani et al. (2016) characterized the genetic variability of uchuva ecotypes from Fraiburgo, Vacaria, Caçador, Lages, Colombia, and Peru and selected promising parents based on fruit traits. Evaluated traits were; husk weight, fruit weight, 1000-seed weight, and fruit diameter. The most distinguishing morphological variables that were identified included fruit diameter, husk weight, and fruit weight. The study demonstrated distinct morphological differences among the *P. peruviana* ecotype populations from Vacaria, Caçador, Lages, and Peru.

Chacón et al. (2016) analyzed 85 uchuva accessions from Colombia using 15 SSR markers. Their results showed two distinguishable groups of accessions related to geography. One group included cultivated and non-cultivated accessions from the eastern Andes (Norte de Santander, Santander, Boyacá, and Cundinamarca) and the other group comprised cultivated and non-cultivated accessions from the central and western Andes (Antioquia, Caldas, Cauca, and Nariño). The low level of genetic diversity within the two distinct groups; however, indicated transport of seeds across neighboring regions is the dominant seed source for propagation and genetic movement.

Kumar et al. (2016) studied the genetic diversity of 12 Indian uchuva genotypes based on morphological parameters. These were grouped into four clusters. Clustering indicated no association between geographical distribution of genotypes or genetic diversity. Fruit diameter was the factor among other traits, such as duration of fruit growth to maturity, number of flowers per branch, fruit weight, and internode length that contributed most to the divergence of genotypes.

Kumar et al. (2018) also examined the genetic diversity of 12 uchuva genotypes from India. They used 20 Random Amplified Polymorphic DNA (RAPD) primers to screen for the potential polymorphic primers for use in evaluating the genetic diversity of uchuva accessions. The least diverged genotypes were CITH Sel-1 and CITH Sel-15. The most divergent genotypes were CITH Sel-7, CITH Sel-9, CITH Sel-16, CITH Sel-5, SS/VK/501, and SS/VK/601. The latter group could, therefore, be used as parents in hybridization programs.

Delgado-Bastidas et al. (2019) evaluated the genetic variability of 40 uchuva genotypes using six SSRs. These six SSR markers were chosen based on their high polymorphism in uchuva. Genotypes were separated into three populations: double-haploid lines, *Fusarium-oxysporum*-tolerant genotypes, and accessions from Universidad de Nariño uchuva germplasm collection. The most informative markers were SSR15 and SSR13. Polymorphic loci were 22.2%, and the average expected heterozygosity was low (0.09) among the uchuva genotypes.

Bonilla et al. (2019) evaluated the genetic diversity of three populations of uchuva from Peru. Protein quantification and polymorphism of seed storage proteins (SSPs) was accomplished by protein denaturation and separation using polyacrylamide gel electrophoresis (SDS-PAGE). Populations were selected from Celendín province (Celendino ecotype), San Pablo province (Agroandino ecotype), and Cajabamba province (Cajabamba ecotype), all in Cajamarca State, Peru. Albumins showed polymorphism revealing 21 proteins between ~6.5 and ~45 kDa and three different electrophoretic profiles among the three populations. The Cajabamba population had the highest genetic diversity followed by the Celendín population. San Pablo's province population was uniform. Furthermore, based on seed protein analyses it is not possible to distinguish the Agroandino, Cajabamba, and Celendino uchuva ecotypes.

11.1 Chromosome Number

Uchuva is a diploid plant with 24 chromosomes ($2n = 24$) (Menzel 1951; Perón et al. 1989; Liberato et al. 2014; Azeez and Faluyi 2019). However, Rodríguez and Bueno (2006) examined the karyotype of five uchuva ecotypes in Colombia, three wild and two cultivated. The three wild ecotypes came from plants growing in Villa de Leyva (Boyacá State) and La Calera and Choachí (both in Cundinamarca State). The two cultivated ecotypes were Colombia, a native ecotype collected from Subachoque (Cundinamarca State) and Kenya, an introduced African ecotype that was collected from Paipa (Boyacá State). The three wild ecotypes were 2n = 24. Colombia's ecotype was 2n = 32, and Kenya's ecotype was 2n = 48. Furthermore, the various ecotypes displayed morphological differences, such as leaf area, stomate density, guard-cell chloroplast number, and fruit diameter. These differences were attributed to ploidy variation among the ecotypes.

Lagos et al. (2005) and Liberato et al. (2014) reported that uchuva ecotypes have ploidy variations between diploid (2n = 24) and tetraploid (2n = 48) in Colombia. Recently, Trevisani et al. (2018) determined the ploidy level of uchuva genotypes grown in southern Brazil. The chromosome number of four observed populations was found to have 48 chromosomes, classifying them as polyploid with tetraploid cells. They, thus, concluded that uchuva populations are tetraploid (2n = 4x = 48) in Colombia and Peru as are populations from the southern region of Brazil.

References

Azeez SO, Faluyi JO (2019) Karyotypic studies of four *Physalis* species from Nigeria. Acta Bot Hungar 61:5–9. https://doi.org/10.1556/034.61.2019.1-2.2

Bonilla M, Espinosa K, Posso A et al (2008) Caracterización molecular de 43 accesiones de uchuva de seis departamentos de Colombia. Acta Agron 57:109–115

Bonilla H, Carbajal Y, Siles M, López A (2019) Genetic diversity in three populations of *Physalis peruviana* using fractionation and electrophoretic patterns of seed storage protein. Rev Peru Biol 26:243–250. https://doi.org/10.15381/rpb.v26i2.16370

Chacón M, Sánchez Y, Barrero L (2016) Genetic structure of a Colombian cape gooseberry (*Physalis peruviana* L.) collection by means of microsatellite markers. Agron Colomb 34:5–16. https://doi.org/10.15446/agron.colomb.v34n1.52960

Delgado-Bastidas N, Lagos-Santander L, Lagos-Burbano T (2019) View of genetic diversity of 40 genotypes of cape gooseberry *Physalis peruviana* L. using microsatellite markers. Rev Cienc Agríc 36:91–103

Enciso-Rodríguez FE, González C, Rodríguez EA et al (2013) Identification of immunity related genes to study the *Physalis peruviana* – *Fusarium oxysporum* Pathosystem. PLoS One 8:e68500. https://doi.org/10.1371/journal.pone.0068500

Garzón-Martínez GA, Osorio-Guarín JA, Delgadillo-Durán P et al (2015) Genetic diversity and population structure in *Physalis peruviana* and related taxa based on InDels and SNPs derived from COSII and IRG markers. Plant Gene 4:29–37. https://doi.org/10.1016/j.plgene.2015.09.003

Kumar V, Sahay S, Ahmad F et al (2016) Genetic divergence of cape gooseberry (*Physalis peruviana* L.) genotypes in India. Int J Agric Environ Biotechnol Cit IJAEB 9:1–4. https://doi.org/1 0.5958/2230-732X.2016.00001.2

Kumar V, Sahay S, Singh RS et al (2018) Molecular marker based genetic diversity analysis in cape gooseberry (*Physalis peruviana* L.). Curr. J Appl Sci Technol 31:1–6

Lagos T, Vallejo F, Caetano C et al (2005) Comportamiento meiótico de algunos genotipos de *Physalis peruviana* L. Fitotec Colomb 5:1–12

Liberato S, Sánchez-Betancourt E, Argüelles J et al (2014) Citogenética de genotipos de uchuva, *Physalis peruviana* L., y *Physalis floridana* Rydb., con respuesta diferencial a *Fusarium oxysporum*. Corpoica Cienc Tecnol Agropecu 15:51–61. https://doi.org/10.21930/rcta.vol15_num1_art:396

Menzel M (1951) The cytotaxonomy and genetics of *Physalis*. Proc Am Philos Soc 95:132–183

Morillo A, Villota D, Lagos T, Ordóñez H (2001) Caracterización morfológica y molecular de 18 introducciones de uchuva *Physalis peruviana* L. de la colección de la Universidad de Nariño. Rev Fac Nac Agric Medellín 64:6043–6053

Morillo-Coronado A, González Castillo J, Morillo Coronado Y (2018) Caracterización de la diversidad genética de uchuva (*Physalis peruviana* l.) En Boyacá. Biotecnol Sect Agropecu Agoindust 16:26–33

Muñoz JE, Morillo AC, Morillo Y (2008) Microsatélites amplificados al azar (RAM) en estudios de diversidad genética vegetal. Acta Agron 57:219–229

Osorio-Guarín JA, Enciso-Rodríguez FE, González C et al (2016) Association analysis for disease resistance to *Fusarium oxysporum* in cape gooseberry (*Physalis peruviana* L). BMC Genomics 17:248. https://doi.org/10.1186/s12864-016-2568-7

Perón J, Demaure E, Hannetel C (1989) Les possibilities d'introduction et de developpement de solanacees et de cucurbitacees d'origine tropicale en France. Acta Hortic 242:179–186

Rodríguez N, Bueno M (2006) Estudio de la diversidad citogenética de *Physalis peruviana* L. (Solanaceae). Acta Biol Colomb 11:75–85

Simbaqueba J, Sánchez P, Sanchez E et al (2011) Development and characterization of microsatellite markers for the cape gooseberry *Physalis peruviana*. PLoS One 6:e26719. https://doi.org/10.1371/journal.pone.0026719

Trevisani N, Schmit R, Beck M et al (2016) Selection of fisális populations for hibridizations, based on fruit traits. Rev Bras Frutic 38:e-568. https://doi.org/10.1590/0100-29452016568

Trevisani N, Melo RC de, Pierre PMO et al (2018) Ploidy and DNA content of cape gooseberry populations grown in southern Brazil. Caryologia 71:414–419. https://doi.org/10.1080/00087114.2018.1494440

Chapter 12
Morphological Diversity

Abstract The morphological diversity of uchuva plants has been investigated in uchuva accessions from different Colombian localities. Both quantitative and qualitative morphological features have been used to study diversity. Phenological diversity has been studied through multivariate statistical analysis programs, such as Principal Component Analysis (PCA) and Multiple Correspondence Analysis (MCA). Cluster analysis has also been used to explore the anatomical diversity of uchuva accessions.

The morphological diversity of uchuva has attracted interest from a number of researchers, particularly, those studies using uchuva accessions from different Colombian localities. For example, Morillo et al. (2011) analyzed 11 quantitative and eight qualitative morphological features of 18 uchuva accessions housed at Universidad de Nariño, Colombia. Quantitative morphological variables were analyzed using a Principal Component Analysis (PCA), a multivariate statistical analysis that explained 81.75% of the total variation in the first three axes. The first axis explained 42.24% of the total quantitative variability and was associated with variables, such as calyx form, fruit weight, and equatorial, and polar diameters of the fruit. Qualitative morphological variables were analyzed though a Multiple Correspondence Analysis (MCA). This analysis explained 61.51% of the variability linked to five factors associated with flower and calyx morphology.

Lagos et al. (2001) characterized 50 uchuva accessions from Nariño State, Colombia. They described 12 morphological variables. Fruit development, production, leaf size, and flower development explained 70% of the variability. Accessions were grouped into five clusters based on fruit weight, fruit diameter, and pulp weight.

Lagos et al. (2003) used 12 quantitative and 28 qualitative morphological features of 50 accessions of uchuva from Nariño, Colombia. The first component of the PCA explained 70% of the total variation. This variation was explained by traits such as number of seeds per fruit, dry weight of seeds per fruit, fruit pulp weight, and mature fruit weight. Furthermore, accessions identified as UN4, UN5, UN14,

F. Ramírez, T. L. Davenport, *Uchuva (Physalis peruviana L.) Reproductive Biology*, https://doi.org/10.1007/978-3-030-66552-4_12

UN16, UN17, UN21, UN22, UN24, UN31, and UN35 had higher weight and fruit size compared to the other observed uchuva accessions.

Trillos González et al. (2008) used 69 morphological traits to differentiate 46 uchuva accessions from AGROSAVIA, La Selva, Antioquia State, Colombia. They concluded that 56 morphological traits provided 81.16% of the certainty in differentiating the accessions.

Bonilla et al. (2008) characterized 24 uchuva accessions growing at Universidad Nacional de Colombia, Palmira, Valle del Cauca State. This work identified 27 morphological traits. They included 10 quantitative and 17 qualitative features. Their results showed that 65.64% of qualitative variability was explained in three axes by MCA. The first axis comprised 38.53% of the qualitative variation and was associated with petal maculation color, anthers color, mature fruit color, and seed color. The first two axes of the PCA explained 32.04% and 17.02% of the quantitative variation among accession traits. Most of the variability was in fruit weight, fruit length and width, seed dry and fresh weights, number of seeds, and soluble solids.

Madriñán Palomino et al. (2011) characterized 29 uchuva accessions from Universidad Nacional de Colombia in Palmira. They used seven qualitative morphological descriptors, such as leaf, fruit, and calyx shape and color of style and pedicel of immature and mature fruit. Ten quantitative morphological features associated with flowers, fruits, and seeds were used. Five clusters were obtained through a dendrogram. The greatest variability was associated with soluble solids, fruit weight with and without calyx, fruit length, and width.

Peña et al. (2011) determined the seed per fruit relationship among Colombia, Kenya, and South African uchuva ecotypes. Morphological variables were fresh weight, equatorial diameter, polar diameter, and number, and weight of seeds. PCA analysis showed that the Colombian ecotype was different from the other two. This ecotype had a lower average fresh weight, fruit polar and equatorial diameter, but a higher seed index (number of seeds per 100 g of fruit).

Twenty three quantitative and seven qualitative morphological traits were used to characterize the geographic origin and biological status (cultivated, wild type, or undetermined) of 54 uchuva accessions from Boyacá State, and the states of Cundinamarca and Santander, Colombia (Herrera et al. 2012). The first two PCA components explained 52.42% of the total variation. The first component of the PCA explained 39.79% of the total variation. This component was positively associated with fruit size and weight and was negatively associated with fruit pH and maturity.

Wild uchuva ecotypes have not been extensively investigated in Peru. Recently, Aguilar (2018) characterized a wild-type ecotype from the Yungay region of Peru using morphological traits such as fruit weight, polar diameter, equatorial diameter, and seed dry weight. Fruit weight ranged from 1.5 to 5.1 g, polar diameter from 14 to 22 mm, equatorial diameter from 14 to 20 mm and seed dry weight from 0.064 to 0.231 g.

References

Aguilar M (2018) Evaluación preliminar de la morfología de frutos *Physalis peruviana* L. ecotipo Yungay (Ancash – Perú). Rev Dr UMH 4:1

Bonilla M, Espinosa K, Posso A et al (2008) Caracterización morfológica de 24 accesiones de uchuva del banco de germoplasma de la Universidad Nacional de Colombia Sede Palmira. Acta Agron 57:101–108

Herrera A, Fischer G, Chacón M (2012) Agronomical evaluation of cape gooseberries (*Physalis peruviana* L.) from central and north-eastern Colombia. Agron Colomb 30:15–24

Lagos T, Criollo H, Ibarra A, Hejeile H (2001) Caracterización Morfológica de la colección Nariño de *Physalis peruviana* L. In: Congreso de la asociación colombiana de Fitomejoramiento y producción de cultivos. Tolima, p 29

Lagos T, Criollo H, Ibarra A, Hejeile H (2003) Caracterización morfológica de la colección Nariño de uvilla o uchuva *Physalis peruviana*. Fitotec Colomb 3:1–9

Madriñán Palomino C, Muñoz Flórez J, Vásquez Amariles H, Barrera Marín N (2011) Caracterización morfológica de 29 introducciones de *Physalis peruviana* L. de la colección de trabajo de la Universidad Nacional de Colombia Sede Palmira. Acta Agron 60:68–75

Morillo A, Villota D, Lagos T, Ordóñez H (2011) Caracterización morfológica y molecular de 18 introducciones de uchuva *Physalis peruviana* L. de la colección de la Universidad de Nariño. Rev Fac Nac Agron Medellín 64:6043–6053

Peña JF, Ayala JD, Fischer G et al (2011) Relaciones semilla-fruto en tres ecotipos de uchuva (*Physalis peruviana* L.). Rev Colomb Cienc Hortíc 4:43–54. https://doi.org/10.17584/rcch.2010v4i1.1224

Trillos González O, Cotes Torres JM, Medina Cano CI et al (2008) Caracterización morfológica de cuarenta y seis accesiones de uchuva (*Physalis peruviana* L.), en Antioquia (Colombia). Rev Bras Frutic 30:708–715. https://doi.org/10.1590/S0100-29452008000300025

Conclusion

This book covered the reproductive biology of uchuva, an Andean member of the solanaceous or nightshade family. The transition from vegetative to reproductive stage is an important physiological event that is not well understood. This transition underlies a number of genetic, biochemical and morphological changes that enable the plant to produce flowers and fruits. The phloem-mobile, Flowering locus T protein (FT), the universal florigenic inducer of flowering, is responsible for the transition of vegetative to flowering buds in many plants, but it remains to be elucidated in uchuva. However, the first bifurcation of the main stem has been observed to be associated with the transition from juvenile vegetative to mature plant bearing flowers and fruits.

We have examined the phyllotaxy of uchuva. The stem architecture consists of alternate inserted main leaves. A newly described spur grows from each lateral branch point, and it is associated with alternate but perpendicular oriented leaves to those of the main stem. The spurs and main stem leaves have a repeated triplet disposition on uchuva stems.

Mature uchuva plants constantly produce buds, leaves, flowers, and fruit throughout the year under tropical conditions (Ramírez et al. 2013). This is similar to other solanaceous fruits such as lulo, cocona, and tree tomato (Ramírez et al. 2018; Ramírez and Kallarackal 2019; Ramírez and Davenport 2020; Ramírez 2020, 2021). This ever-bearing habit, which produces fruits through the year is beneficial for uchuva growers and consumers. In contrast to constantly warm tropical environments, uchuva flowering and fruiting events occur seasonally at particular times in sub-tropical or temperate climates.

Phenological scales such as the German BBCH scale have been used to characterize the growth and development of uchuva leaves, flowers and fruits. This scale can be used to establish horticultural management practices for plant fertilization, irrigation, pruning, and disease control.

Pollination appears to be closely associated with insect visitation; however, studies are lacking regarding the actual role of insects in pollination vs pollen deposition

F. Ramírez, T. L. Davenport, *Uchuva (Physalis peruviana L.) Reproductive
Biology*, https://doi.org/10.1007/978-3-030-66552-4

by wind. Although insects have been claimed to be the main pollination mechanism for fruit set under tropical conditions, wind has been largely unexplored for pollen transfer from one flower to another. The role of wind needs to be resolved. Most investigations of insect visitations on uchuva have been conducted in the tropics and within its area of origin. Fewer investigations have focused on insect activity in uchuva flowers growing in sub-tropical conditions with marked annual variation in temperature and precipitation.

Multiple arthropods have been observed visiting the flowers of uchuva in the tropics. These include honeybees, bumblebees, stingless bees, wasps, beetles, hemipterans, and ants. Particularly, bumble bees, such as *Bombus atratus*, make short visits to the flowers of uchuva. Native Andean stingless bees have also been observed visiting uchuva flowers for long periods of time.

Future studies should focus more on pollen morphology, stigma receptivity, and anther dehiscence. These topics require more research among uchuva accessions. Research derived from the precise timing of anther dehiscence and stigma receptivity is required to better understand the dynamics and best conditions for effective pollination. Pollen viability and germination also require more research. This is particularly important to establish pollen storage conditions. Moreover, self pollination is not well understood. Few studies have evaluated the role of wind in uchuva pollination. Caging experiments excluding insects, coupled with molecular techniques are necessary to understand the role of cross and self pollination mediated by wind. Paternal molecular techniques, e.g. SSR markers could be applied to understand cross pollination between uchuva accessions.

Uchuva seed propagation is the most widely used method for commercial purposes, especially in developing countries, where most uchuva plants are grown. This is due to the ease of propagation and high germination rate, which is over 85%. Furthermore, plants propagated by seed adapt well in the field. Fruit soluble solids are higher in seed-propagated plants than in those propagated from cuttings.

Uchuva propagation by cuttings is a technique to obtain plant clones from a mother plant. This asexual propagation technique produces uchuva fruits earlier than those obtained from seed. Propagation by cuttings is not used commercially due to several drawbacks, such as high cost, poorly developed root systems, susceptibility to wind damage, and reduced fruit quality. Plants propagated by cuttings are also more susceptible to fruit splitting in contrast to seed-propagated plants; however, plants propagated by cuttings yielded more fruit in New Zealand compared to plants grown from seed.

Throughout the world, a number of ecotypes have been identified due to their high productivity, good fruit quality, and nutritious properties that benefit consumers. Ecotype diversity is key for adapting to different environmental conditions. Furthermore, this diversity can be used to select for resistance to diseases and for plant quality improvement. People prefer some ecotypes over others, and utilization of these preferences are essential to expand uchuva demand and marketability. A clear understanding of the reproductive biology of uchuva is vital to achieve these goals. Traditional selection of the best plant material from cuttings or seeds has been the main approach to improve ecotypes.

In vitro culture techniques are also being developed to obtain plant material suitable for improved horticultural production. *In vitro* propagation techniques have used plant growth regulators to stimulate root and shoot growth and callus formation. These include the auxins, indole-3-acetic acid (IAA), indole-3-butyric acid (IBA), naphthalene acetic acid (NAA), 2,4-dichlorophenoxyacetic acid (2,4-D), the cytokinins, 6-benzylaminopurine (6-BAP) and kinetin, and the gibberellin, gibberellic acid 3 (GA$_3$). Most *in vitro* propagation studies focused on the initial development of roots, shoots, and/or callus formation; however, they have not measured subsequent uchuva plant growth, fruit growth, development, or yield after transplant to the field or in greenhouse conditions. These are key events to evaluate the success of *in vitro* propagation.

The calyx plays an important function during fruit development. This is due to its protective and carbohydrate translocation functions during fruit growth and development. Husk color change is a reliable indicator of fruit maturity. Skin color of the fruit is also a reliable indicator of horticultural maturity in uchuva. Several scales have used fruit color or husk color to determine harvest maturity. These scales are important for production purposes and are based on nondestructive methodologies to judge fruit maturity. Furthermore, recent technological improvements, such as fruit maturity sensors, have been developed as reliable tools to measure the physiological maturity at harvest.

The uchuva fruit is a source of compounds with potential health benefits for human beings. Particularly, it is a rich source of provitamin A, vitamin B complex (niacin, thiamin and B$_{12}$), vitamin C, and β-carotene. Vitamin C protects the human body against heart-related diseases, cancer, improves immune system function, and collagen biosynthesis. Vitamin B has a role in carbohydrate metabolism by converting sugars to energy, and aiding in fat and protein metabolism. β-carotene plays an important role in bone development, vision, cell division and differentiation, and reproduction. Uchuva has antispasmodic, antiseptic, sedative, diuretic, and analgesic medicinal properties. It has been used in traditional medicine to treat cancer, hepatitis, asthma, malaria, and dermatitis, but these have not been scientifically authenticated.

Germplasm repositories play an essential role in perpetuating accession diversity, breeding and improvement of plant traits. Colombian germplasm repositories, research institutes, and universities have played an important role in establishing uchuva breeding programs seeking increased yield and disease resistance. For example, disease resistant accessions have been selected against *F. oxysporum*, which generates high economic loses in Colombia. In Ecuador and Peru, germplasm repositories have focused more on traditional uchuva ecotypes. These could improve yield and disease resistance, but more research is warranted. Uchuva hybrids have been difficult to obtain and in some cases unsuccessful. For example, in a cross between *P. angulata* x *P. pubescens*, their hybrid produced few flower buds, and no mature fruits were harvested. Successful hybrids have been obtained from crosses between *P. minima* L. and *P. peruviana* by *in vitro* culture techniques using an MS medium supplemented with cytokinin and auxin. The obtained hybrid produced viable fruit and offspring.

It is possible that uchuva has an interspecific incompatibility system as evidenced by pollen meiotic irregularities and chromosome aberrations occurring in some hybrids during the first generation. In other interspecific crosses, the incompatibility system operates during the first generation failing to produce viable offspring. A self-incompatibility system has not been identified in uchuva. Menzel (1951) reported that several *Physalis* species (Solanaceae), have no self-incompatibility; however, it was observed that bagged flowers of *P. ixocarpa* Brot. (a cross between *P. aeguata* Jacq. and *P. capscicifolia* Rydb.) and *P. viscosa*, failed to set fruit. Other crossed *Physalis* species were able to set abundant fruits with normal or near normal number of seeds. Furthermore, Pandey (1957) reported that *Physalis* has only the one-locus, multi-allelic gametophytic system of self-incompatibility.

Uchuva genetic diversity has been studied using different techniques, such as Random Amplified Microsatellites (RAMs), Simple Sequence Repeat Markers (SSR), seed storage proteins (SSPs), Conserved Orthologous Sequence (COSII), and Immunity Related Gene (IRGs) protocols. High genetic variability has been reported for uchuva accessions from Valle del Cauca, Colombia.

Qualitative and quantitative traits have been used to characterize the morphological diversity of uchuva accessions in Colombia and Peru. This is important for selecting best marketable fruit quality traits, such as pulp weight, polar and equatorial diameter, and number of seeds.

Uchuva is a wonderful fruit that needs to be exported to other countries so they can appreciate its delicious quality. It has a delightful flavor and is easily eaten out of hand like cherry tomatoes with a distinctive tart, fruity flavor. Uchuva is a fruit that can also be used in a number of dishes. Its bright, yellow color enhances the presentation and flavor of gourmet foods. This marvelous fruit can only be exported if the fruit is grown in a certified fruit-fly free zone. Due to the presence of *Anastrepha obliqua* Macquart, a fruit fly that plagues orchards in Colombia, additional fruit-fly free zones could be established using effective insect control protocols in those areas. Uchuva fruit has a number of health-related benefits mentioned above, which include nutritional properties and molecules that may be used in disease prevention and cure. This book consolidates scientific viewpoints about the reproductive biology of uchuva from researchers around the world. It also has taken into account the research carried out by the lead author in research on uchuva and related solanaceous plants. We hope this book stimulates new and innovative research in the field of uchuva reproductive biology.

References

Menzel M (1951) The cytotaxonomy and genetics of *Physalis*. Proc Am Philos Soc 95:132–183

Pandey KK (1957) Genetics of self-incompatibility in *Physalis ixocarpa* Brot. – a new system. Am J Bot 10:879–887. https://doi.org/10.2307/2438909

Ramírez F (2020) Cocona (*Solanum sessiliflorum* Dunal) reproductive physiology: a review. Genet Resour Crop Evol 67:293–311

Ramírez F (2021) Notes about Lulo (*Solanum quitoense* Lam.): an important South American under-utilized plant. Genet Resour Crop Evol 68:93–100. https://doi.org/10.1007/s10722-020-01059-3

Ramírez F, Davenport TL (2020) The development of lulo plants (*Solanum quitoense* Lam. var. *septentrionale*) characterized by BBCH and Landmark phenological scales. Int J Fruit Sci 20:562–585. https://doi.org/10.1080/15538362.2019.1613470

Ramírez F, Kallarackal J (2019) Tree tomato (*Solanum betaceum* Cav.) reproductive physiology: a review. Sci Hortic (Amsterdam) 248:206–215

Ramírez F, Fischer G, Davenport TL et al (2013) Cape gooseberry (*Physalis peruviana* L.) phenology according to the BBCH phenological scale. Sci Hortic (Amsterdam) 162:39–42. https://doi.org/10.1016/j.scienta.2013.07.033

Ramírez F, Kallarackal J, Davenport TL (2018) Lulo (*Solanum quitoense* Lam.) reproductive physiology: a review. Sci Hortic (Amsterdam) 238:163–176. https://doi.org/10.1016/j.scienta.2018.04.046

F. Ramírez, T. L. Davenport, *Uchuva (Physalis peruviana L.) Reproductive Biology*, https://doi.org/10.1007/978-3-030-66552-4

Printed in the United States
by Baker & Taylor Publisher Services